# THE DISAPPEARING ISLANDS OF THE CHESAPEAKE

# CHESAPEAKE BAY

Garrett
Maids · Fishing Battery
Spesutie

(Spry) · Pooles
Hart & Miller
Fort Carroll

Caccaway
Dobbins · Gibson
St Helena · Eastern Neck

Kent · Parsons
Bodkin · Wye
(Three Sisters) · Bruffs

Poplar

Tilghman

(Sharps)·

James

Taylors

Broomes

Solomons · Barren

Cobb

St Clements · Hoopers

Tippety Witchety
Bloodsworth · Deal
Holland · South Marsh

Smith · Janes

Cedar

Great & Little
Fox
Tangier

Watts

Gwynn's

Jamestown

## SCALE

### NAUTICAL MILES
0   5   10   15   20   25

0   10   20   30   40
### KILOMETERS

# THE DISAPPEARING ISLANDS OF THE CHESAPEAKE

WILLIAM B. CRONIN

THE JOHNS HOPKINS UNIVERSITY PRESS · BALTIMORE AND LONDON

IN ASSOCIATION WITH THE CALVERT MARINE MUSEUM · CHESAPEAKE BAY MARITIME MUSEUM ·
MARINERS' MUSEUM · MARYLAND HISTORICAL SOCIETY

The Johns Hopkins University Press

2715 North Charles Street

Baltimore, Maryland 21218-4363

www.press.jhu.edu

Cronin, William B.

  The disappearing islands of the Chesapeake / William B. Cronin.

    p.  cm.

  Includes bibliographical references and index.

  ISBN 0-8018-7435-1 (alk. paper)

1. Islands—Chesapeake Bay (Md. and Va.)—History. 2. Chesapeake Bay (Md. and

Va.)—History. 3. Chesapeake Bay (Md. and Va.)—Description and travel. 4. Chesapeake

Bay (Md. and Va.)—Environmental conditions. 5. Island ecology—Chesapeake Bay

(Md. and Va.) I. Calvert Marine Museum. II. Chesapeake Bay Maritime Museum.

III. Mariners' Museum (Newport News, Va.) IV. Maryland Historical Society. V. Title.

  F187.C5C76 2005

  551.46′1347—dc22

2004015972

A catalog record for this book is available from the British Library.

*Frontispiece:* Chesapeake Bay and its islands in the early twenty-first century. (Bill Nelson, after author's sketch)

# CONTENTS

## PREFACE

This book is the culmination of my thirty-year exploration of the Chesapeake Bay and its islands, whose number and variety continually amaze me. Before retiring as a staff oceanographer for the Johns Hopkins University's Chesapeake Bay Institute and captain of the research vessels *Lydia Louise I* and *II,* I covered much of the bay and visited several island ports. Often on these cruises on the Chesapeake, the Continental Shelf, and many bay tributaries, my guide and inspiration was Dr. Jerry Schubel. The best cruise of my life was the summer we spent "banging the bottom," taking seismological soundings of bay sediments aboard the *Lydia Louise II.*

I was also fortunate to visit many of the bay's islands in my sailboat, *Ginger.* The twenty-five-foot Coronado was almost as much of a companion as the beloved Chesapeake Bay retriever whose name she bore. Alone and with friends and colleagues, I sailed about, frequently dropping anchor in sheltered coves and off sandy bay beaches or wide salt marshes to photograph windswept bluffs, crumbling bulkheads, ramshackle houses, and lonely graveyards on Chesapeake islands. Sometimes I went ashore and on occasion found islanders who told me of their lives. More often an island's inhabitants were waterfowl, song birds, small animals, and deer.

Over the years I returned to several islands. I'd first gone to Poplar Island in the 1970s to look for arrowheads. There were few traces of human habitation, and the surrounding waters were eating away at the land, loosening the soil around the roots of trees and toppling them into the water. Chief among the island's occupants were a sizeable herd of deer and large flocks of ospreys, herons, and egrets. When I last explored Poplar Island twenty years later, only its highest points remained, forming six small islands with few surviving trees and scattered deer tracks to suggest that some were left.

By that time, I'd begun writing a series of articles about the islands for *Chesapeake Bay Magazine* and made it a point to visit those that I wrote about and to document them with photographs. As often as possible, I located people who were familiar with a particular island and took them in my boat to help me see the island as they remembered it. My photographic expeditions also took me into the skies, thanks to a friend and retired professional pilot who willingly flew me over the bay and its islands. My interest didn't end with the magazine series. As I added islands, I broadened my research, searching for old charts, maps, and information that would help me tell their stories. I scoured libraries and other resources at the Chesapeake Bay Maritime Museum and the Mariners' Museum, among others; at state and county historical societies; at the Maryland State Archives; and at the National Archives.

Back home, I began writing, drawing upon my own knowledge and observations and the materials that I had gathered. As exciting as my actual explorations, the research took me back to the adventurous men and women who were the earliest settlers of Maryland and Virginia. Many of the islands shared in the same historic events, and their stories, individually and collectively, tell about both good and troubled times.

Often my explorations were solitary, but no one is truly alone in such endeavors. In long talks, flights over the bay, photographic expeditions on land and afloat, and searches for archival materials, many people assisted me before I could begin to put it all in writing. I am particularly indebted to Dr. Schubel, who was a source of superb inspiration and guidance, personal and professional, and to the staff of the Chesapeake Bay Institute, especially the late Dr. D. W. Pritchard, the director, and the early staff, including Wayne Burt, Jim Carpenter, Dave Carritt, Tom Hopkins, Blair Kinsman, James McGarry, Martin Pollack, Bud Whaley, Dick Whaley, and Jerry Williams. The expertise of the research vessel captains and crews, Dave Booth, Winnie Booth, Norman Gilbert, "Stubby" Gilbert, Bill Harris, and Charles Wessels, made research easier for the scientists aboard. Visiting scientists regularly shared their knowledge, setting an example and stimulating further investigation.

Many others helped in various ways. Bill Tinkler, a professional pilot, volunteered his high-wing plane whenever I needed a new set of aerial photographs. Those who shared personal recollections of their particular islands included Vernon Bradshaw, Dr. Eugene Cronin, Richard Earle, Wilson Ford, Frank Hammond, John Hannon, Tom Horton, Peter Jay, Eric Klinglehofer, Ella May Lewald, Margaret Lewis, Michael Macielag, Madison Mitchell, Nat and Lola Parks, Chan Rippon, Patricia Rodgers, Gregory Stiverson, Joe Ward, and many, many others whose names escape my memory. Professional librarians helped me locate information, particularly Mary Hanning of the Anne Arundel County Public Library. All of them were gracious and helpful. I also wish to thank Connie Lewis of the Maryland Department of Natural Resources for the latest fish statistics.

I am indebted to Lee Nelson who carefully edited my articles for *Chesapeake Bay Magazine*. I also wish to express my appreciation for the efforts of my copyeditor, Ann Jensen, who corrected my prose, spelling errors, improper dates, and misquoted histories and made this book more informative and interesting from her own personal knowledge.

I deeply appreciate the cooperation of my son, Dr. Thomas W. Cronin, 2004 Teacher of the Year at the University of Maryland Baltimore Campus, whose computer expertise clarified many archival charts and photographs and whose support and encouragement helped me through many problems.

Finally, my wife Elizabeth's constant encouragement inspired me beyond measure.

# A NOTE ON THE PHOTOGRAPHY OF A. AUBREY BODINE (1906–1970)

A true son of the Chesapeake, A. Aubrey Bodine was born in Baltimore in 1906. At the tender age of fourteen, he began working as a messenger boy for the *Baltimore Sun,* earning eight dollars a week. The relationship lasted half a century. Keenly interested in photography, the young Bodine often accompanied the newspaper's commercial photographer, Herbert Moore, on assignments, carrying his tripod camera and heavy glass plates. One day while Moore prepared to expose a picture, the flash powder exploded in his face, hospitalizing him with bad burns. Bodine filled in, doing so well that the *Baltimore Sun* promoted him to commercial photographer. Three years later, in 1927, he was promoted to the post of *Sunday Sun* photographer.

He soon enjoyed a wide reputation, successfully submitting his photographs to salon exhibitions all over the country. In addition to his work for the *Sun,* Bodine accepted commercial assignments that ranged from personal portraits to images of steel mills. His photographs appeared in magazines with national and even international audiences, including *Harper's Bazaar* and *Life.* He contributed articles on a variety of subjects to photography magazines and lectured—in his own fashion—to numerous camera clubs around the country. "He was the best photographer in America," remarked a Philadelphian who often invited Bodine to make appearances, "and near its worst speaker."

Driven by a passion for his work, Bodine once said that he made some of his best photographs when most other photographers were asleep. Perhaps so, but his success also relied on artful manipulation of the photographic process. Bodine subscribed to "pictorialism," a movement in photography that flourished from the 1880s until about the 1920s. It grew out of a felt need on the part of some photographers—at a time when photography seemed synonymous with commercial hacks and Kodak-clicking amateurs—to differentiate their own work as creative art of the highest order. The pictorialist approach featured romantic subject matter, care in composition, and soft focus. A master of composition and focus, Bodine well illustrated a pictorialist practice—images manipulated by means of special techniques in the handling of negatives and printing of glossies. Skilled in the darkroom, he would often shoot separate negatives and then combine them in printing, adding clouds, snow, rain, and even seagulls to his photographs. Though photographic paper is dated to ensure quality, Bodine would intentionally use outdated paper, claiming it produced better exhibition prints. He lengthened the recommended developer times for film. He developed his negatives by inspection, removing them from the chemicals and viewing them under the safe light so that he could have complete control over the process. The ideal photographic print for Bodine was "one in which there is a perfect range of tones, from pitch black areas among the darker shadow, through the middle tones, to the highlights, which should have detail." His fine control of light and artful darkroom technique became the hallmarks of his work.

For most of his outdoor shooting Bodine used his favorite camera, a Linhof. Its large negative—measuring five by seven inches—was ideal for retouching. He kept most of his equipment in his car trunk, and indeed the size of the trunk often determined which new car he bought. Besides cameras and tripods, he carried compass, ax, machete, shovel, bee smoker, a child's white parasol, and toilet paper. The machete, ax, and shovel allowed him to cut trees and branches or uproot fence posts that were in the way of a good shot. The compass assisted in figuring future light when he was caught in strange territory without sunlight. The bee smoker could create mist, while the parasol and toilet paper diffused and softened light from the sun or his flash. Bodine's exquisite photographs soon distinguished him from other photographers of the day.

Whether photographing watermen, cypress trees, or a Maryland power plant, Bodine conveyed a deep love of the scenery and spirit of the Chesapeake Bay. His photographs thus had persuasive power; they encouraged among viewers an awareness and keen appreciation of the uniqueness of the bay and the life around it. Bodine eventually produced three books celebrating the region. *My Maryland* (1952), featuring 174 photographs that he selected from 25,000 negatives, won honors from the

National Lithographers Association as the best printed book for that year. By 1971, it had gone through four printings and sold more than 9,000 copies. *Chesapeake Bay and Tidewater,* first published in 1954, became Bodine's most popular work, selling more than 22,000 copies in five printings and two revised editions. A third book, *The Face of Maryland* (1961), sold 12,000 copies. Yet another, *The Face of Virginia* (1963), sold 9,000 copies by 1971.

Much of Bodine's work reflected his love affair with the islands of the Chesapeake Bay, an engagement that seems to date from a trip he made to the Eastern Shore in 1929 to photograph work boat races. That year he visited Gwynn's Island, Virginia, and Crisfield, Maryland, a ferry link to Smith Island (which he seems to have photographed only later, in 1945 and the early 1950s). He did work at Solomons Island in 1937 and revisited Gwynn's Island two years later and twice more before 1942. In 1939 he also set up his camera on Pooles and Gibson islands, shooting Gibson again in 1941. After World War II, besides several times photographing Smith Island, Bodine worked on Tilghman, Tangier, Great and Little Fox, St. Clement's, and Taylors islands (all except Tangier in Virginia). In 1961 he went again to Deal, and in 1963, apparently making his last working visit to a Chesapeake community out in the bay, he returned to Tilghman Island.

These remote sites fascinated Bodine, who marveled not only at their rich visual material but also at the way of life they sustained. "I am often asked how I find certain subjects," he wrote in the introduction to *Chesapeake Bay and Tidewater.* "Some I get from reading old volumes—one of my hobbies is collecting and reading old books about the Bay country—but most of it comes from talking to people. I enjoy talking to people gathered in country stores, watermen standing around the wharves, country newspaper editors, country agents and the like. They are wonderful people to know, and they know just about everything that's going on." He declared that he could not imagine a "bigger thrill" than making "a picture of dredge boats moving over an oyster bed on a beautiful autumn day. That's the sight," he said, "I hope I can keep photographing for the next 25 years." This interest naturally led him to the islands of the bay and their residents. He admired and respected watermen for their hard work and simple way of life. His photographs combined the pictorial style and his vision of men working in harmony with nature. "One of the most picturesque spots in the country," he wrote in *The Face of Maryland,* had to be Hoopers Island, whose crabbing and fishing captivated him and the spirit of whose natives he found eminently American. "Ask a waterman how things are going and he will simply reply, 'Had a slim day.' But I have noticed that watermen who work regularly can make a comfortable living. Most of them own their own boats, homes and cars. And, above all, they are supremely independent, and I cannot help but envy this independence and their way of life." From capturing scenes of watermen at work to occasional portraits of distinctive personalities, Bodine photographed with increasing ease. Aware of his regional reputation, locals began to recognize and welcome him to the islands.

In 1960, Bodine's photographs traveled with an early, if not the first, exhibition showcasing the work of distinguished photographers from many countries. When the show reached Moscow, a Soviet reviewer praised the photographers; they were "united by a single wish—to show life without embellishments, such as it is in all its variety, with all its joys and difficulties." In the exhibit, Bodine's "monumental image" of a Baltimore stevedore won a silver medal. The judges might as easily have selected a photograph of some hardworking Chesapeake watermen—though one has to wonder how, at the time, their fierce individualism would have played in Moscow. At Bodine's death Jonathan Yardley, writing in the *Washington Post,* summed up his reputation by simply saying that he was "a photographer of quite amazing gifts and accomplishments."

Bodine's photographic legacy now lies in the collections of some twenty nationally renowned museums, libraries, and archives—among them the George Eastman House in Rochester, New York; the Center for Creative Photography in Tucson, Arizona; the J. Paul Getty Museum in Los Angeles; the Maryland Historical Society in Baltimore; and the Mariners' Museum in Newport News, Virginia. Bodine established a working relationship with the Mariners' Museum in 1949, when he served as a juror in the museum's first annual Salon of Marine Photography. This special tie lasted until his death in 1970, and in 1971 Bodine's widow gave the museum 3,200 original glass-plate and film negatives from the master's oeuvre. Including some 4,000 Bodine photographs, this collection depicting life and work on the mid-twentieth-century Chesapeake Bay stands as one of the country's strongest. The Mariners' Museum continues to incorporate Bodine's work into various permanent, temporary, traveling, and on-line exhibitions, visually telling thousands of visitors the history of the bay and its people.

Along with dozens of images from private sources, the Bodine photographs in this book offer merely a sampling of the rich Bodine resources available at the Mariners' Museum and one of its collegial institutions, the Maryland Historical Society. We hope that Mr. Cronin's book attracts serious research in our collections—and stimulates public interest in the bay as did Bodine's original work.

THOMAS MOORE
Curator of Photography
The Mariners' Museum

# A NOTE ON REFERENCES AND ABBREVIATIONS

The following entries, arranged alphabetically by island, include two kinds of source material that any student of the islands of the Chesapeake will find helpful in further research—historic maps and charts and scholarly and popular historical accounts. Works frequently mentioned carry the following abbreviations:

## HISTORIC MAPS AND CHARTS

HC, USCGS   U.S. Coast and Geodetic Survey, Department of Commerce, hydrographic charts. The federal government has issued them since 1816 and stores them in a distribution branch, N/CG23, National Ocean Service, National Oceanographic and Atmospheric Administration, Riverdale, Maryland 20737.

Lake et al., *Atlas*   Lake, Griffing, and Stevenson. *Atlas* (1877) covering Cecil, Dorchester, Kent, Queen Anne's, Somerset, Talbot, Wicomico, and Worcester Counties.

MdDNR   Maryland Department of Natural Resources photographic surveys of wetlands and submerged aquatic vegetation, along with shore erosion-control revetment studies and plans. They are available for viewing at the DNR office at 580 Taylor Avenue, Annapolis, Maryland 21401.

MdGS   Maryland Geological Survey erosion maps. One may order them from the survey offices at 2300 St. Paul Street, Baltimore, Maryland 21218.

MdIHP   Maryland Inventory of Historic Properties. This survey of Maryland historic sites was made for the Maryland Historical Trust, and the inventory is available in the trust's library for public use. Their offices are located at 100 Community Place, Crownsville, Maryland 21032.

NOAA   National Oceanographic and Atmospheric Administration (NOAA), Department of Commerce, charts. One can obtain them through the agency and also many nautical retail outlets.

Q, USGS   U.S. Geological Survey, Department of the Interior, topographic maps identified as quadrangles, or "quads." They are available from the USGS distribution center, 1200 South Eads Street, Arlington, Virginia 22202, and from private distributors.

TC, USCGS   U.S. Coast and Geodetic Survey, topographic charts (dates vary), stored with the National Archives and Records Administration, College Park, Maryland 20740.

## GENERAL REFERENCES

Arnett et al.   Arnett, Earl, Robert J. Brugger, and Edward C. Papenfuse. *Maryland: A New Guide to the Old Line State*. 2d ed. Baltimore: Johns Hopkins University Press, 1999.

*CBMag*   *Chesapeake Bay Magazine*

deGast    deGast, Robert. *Lighthouses of the Chesapeake*. Baltimore: Johns Hopkins University Press, 1973, 1993.

Eller    Eller, Ernest M. *Chesapeake Bay in the American Revolution*. Centreville, Md.: Tidewater Press, 1981.

Hornberger/ Turbeyville    Hornberger, Patrick, and Linda Turbeyville. *Forgotten Beacons: The Lost Lighthouses of the Chesapeake Bay*. Annapolis: Eastwind Publishing, 1997.

*MdHM*    *Maryland Historical Magazine*

MdHS    Maryland Historical Society

MdHT    Maryland Historical Trust

Mowbray    Mowbray, Calvin W. *Dorchester County Fact Book*. Dorchester Co., Md.: privately printed, 1980.

Papenfuse/ Coale    Papenfuse, Edward C., and Joseph M. Coale III. *The Maryland State Archives Atlas of Historical Maps of Maryland, 1608–1908*. Baltimore: Johns Hopkins University Press, 2003.

Papenfuse et al.    Papenfuse, Edward C., et al., eds. *A Biographical Dictionary of the Maryland Legislature, 1635–1789*. 2 vols. Baltimore: Johns Hopkins University Press, 1979, 1985.

*Sun*    *Baltimore Sun*

*Sun Mag*    *Baltimore Sun Magazine*. Published Sundays.

THE DISAPPEARING ISLANDS OF THE CHESAPEAKE

INTRODUCTION

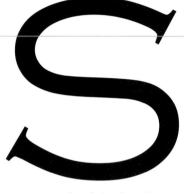

**S**ince Capt. John Smith explored the bay in 1608, men and women of power, wealth, and individual enterprise have linked islands in the Chesapeake with events that shaped the history of the region, the nation, and the world. Travel by water became second nature to those who lived along the bay and its tributaries, which meant that an island could be as desirable as any other piece of land for those who owned boats. Since the first colonists came ashore on Jamestown, Kent, and St. Clement's Islands, Marylanders and Virginians have put island properties to a wide variety of uses. Bay and river islands offered pasturage for livestock, land for nature preserves and agricultural experiments, and in many cases, an ideal spot for a self-contained community of farmers and watermen. Chesapeake islands served as retreats for presidents, senators, congressmen, and a host of lesser officials and politicians. They nurtured inventors, educators, writers, artists, and scholars and aroused equal wonder and despair in naturalists and environmentalists. Many owners, as well as would-be entrepreneurs and the occasional pirate, outlaw, and scoundrel, have been willing to risk much, if not all, on island dreams of fortune or fame tied to such diverse ventures as foreign intrigue, buried treasure, exclusive resorts, Sitka deer, black cats, silk worms, peach brandy, spices, gambling, brothels, and world championship boxing. The real story of the islands, however, emerges in the lives of the thousands of hardworking watermen and farmers and their families who peopled this precarious, water-wrapped world, clinging to their homes, their enterprises, and their independence with tenacity, courage, and, at times, great ingenuity.

The story of Holland Island suggests the course that nature and man have charted for these ragged bits of land. Its history is instructive for any and all who live near the water's edge. In her 1994 master's thesis for the University of Maryland, Sheila Jane Arenstam recorded the effect of rising sea level and how it turned island residents into "environmental refugees."

For nearly three-quarters of a century, a community flourished on Holland Island in Dorchester County on the Eastern Shore. Island men worked the surrounding waters, dredging for oysters in winter and crabbing during the summer. Families raised much of their own food, with enough left over in good years to send to mainland markets. A few kept stores that supplied islanders' daily needs. Holland Island homes were well built and well kept. In 1900 the

*Previous pages:* A tranquil dock house on Gwynn's Island in the mid-twentieth century. (Mariners' Museum, Newport News, Va.)

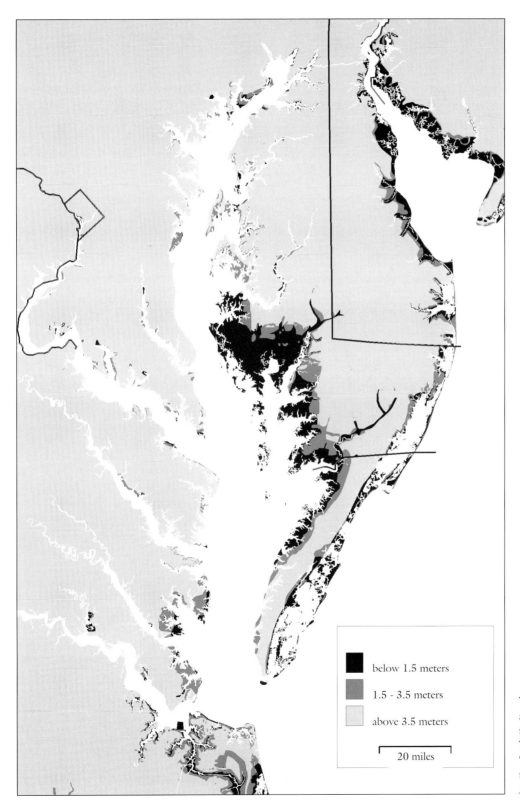

below 1.5 meters

1.5 - 3.5 meters

above 3.5 meters

20 miles

The likely loss of wetlands and islands by the end of the twenty-first century (James Titus and Charles Richman, "Maps of Lands Vulnerable to Sea-Level Rise," *Climate Research* 18 [2001]: 225).

island had its own post office and postmistress, a church and resident preacher, a doctor, and a midwife who brought many a new islander into the world. More than a hundred children attended the Holland Island school. They and their parents gathered for potluck suppers in the community center and cheered their baseball team during games on the diamond at Eagle Point. Holland Island was home, and few, if any, of its 250 residents gave a thought to life there coming to an end. They could survive just about anything, from high tides and wind storms to winter ice. But Holland Island couldn't.

Year after year, northwest winds whipped up thirty miles of open water to hurl foot-high waves at the island's exposed western ridge. Foot by foot, yard by yard, the shore gave way, and by 1900 erosion threatened the homes and businesses strung out along the ridge. Islanders could only watch as the bay ate away at the narrowing strip of land lying between their houses and the water tearing at Holland Island's edges.

Families clung to their homes as long as they could, shoring up a bit here, filling a bit there, but the forces at work were beyond their control. With homes on the verge of falling into the bay or being consumed by spreading salt marsh, people packed up and left for the mainland, taking everything they owned, including their houses. One after another, businesses closed. By 1910 the preacher, postmistress, and doctor were gone. Refusing to let them fall to the elements, the islanders tore down their church and school buildings to be reassembled in mainland towns. Finally in 1918 a summer gale drove the last family from Holland Island. By 1942 the entire western ridge had disappeared, leaving the remainder of the island to flocks of songbirds, blue herons, and sea gulls.

On the eve of the twenty-first century, all that lay above water was the island's eastern ridge where a lone house kept a silent vigil overlooking Holland Straits and the mainland beyond. To the south, the encroaching waters had not quite erased the outline of the baseball diamond, and at the center of the island was the cemetery, its white tombstones stark in a watery field of marsh grass.

Holland Island's settlement, abandonment, and gradual disappearance does not, sad to say, represent a unique occurrence in the history of the Chesapeake. The story has been repeated in one way or another many times. Such stories are often all that remains of many of the Chesapeake's islands, which continue to disappear at an alarming rate. As history tells us, more must certainly follow.

We need only look at the origins of the Chesapeake Bay to see the forces at work on these, its most fragile land masses. This largest of North American estuaries began as the Susquehanna River, which flowed from what is now New York to the Atlantic Ocean. When global warming ended the last Ice Age some fifteen thousand years ago, it melted the glaciers that covered much of North America, causing the sea level to rise and flood the lower Susquehanna valley. Over the next seven thousand years, the waters spread northward, inundating the flat coastal plain of Maryland and Virginia and forming a great bay, which we know today as the Chesapeake.

Fresh water still flows into the bay from the Susquehanna and a number of other rivers as well as thousands of streams and creeks. The estuary's watershed extends into six states to cover a 64,000-square-mile area. Its 8,100 miles of exposed island and mainland shoreline are perpetually vulnerable to the forces of wind and water that have made change a constant in the bay's history.

Over the millennia, as ocean water drowned the Susquehanna River and flowed over the surrounding land, it created a ragged chain of islands strung along the edges of the bay and its larger rivers. Most of the islands lay along the Eastern Shore where they were surrounded by a multitude of shallow channels and vast new salt marshes. Upland vegetation died, trees clawed for a hold as water washed earth from around their roots, finally exposing and killing them with a steady flow of saltwater. Thriving in saltwater, *Spartina* grass took over and spread inland as the sea level rose. This process has never ceased. As the sea rises, the Chesapeake continues to evolve and consume land as it always has.

Today, however, there is a significant and disturbing difference. In the "normal" course of millennial change, the earth should be cooling and heading slowly for another ice age and, among other things, a slowing of the rate of sea level rise. Until 1900, the bay's waters rose at a slow but steady rate of three feet every thousand years, but that has changed in the last one hundred years during which the sea level has already risen one foot. Many scientists believe that this is a sign of the effects of global warming and worry that there could be another two- to three-foot rise over the next century, posing serious problems for bay coastlines and ecosystems. Nowhere is this change more clearly and ominously evident than in the rate at which the Chesapeake's islands are vanishing.

Rising water is not the only force at work on the islands. They also lie exposed to the fury of storms that regularly hit the bay. The sustained forty-mile-per-hour winds of a typical March northeaster will generate high tides and waves up to three feet that tear away at an island's shore. Even more damaging are the hurricanes of late summer and early autumn. Creating a storm surge of as much as seven feet above a normal tide, a hurricane can rapidly and dramatically erode an island's shoreline with the combined force of its high winds, heavy rains, and powerful waves. Beyond storms, waves generated by the prevailing northwest winds that angle across long stretches of the bay in early spring are an additional factor in the rapid erosion rate experienced by islands along the Eastern Shore.

The extent of island erosion was a public concern at least as early as 1914. Working under the auspices of the Maryland Geological Survey, J. F. Hunter reported the amount and rate of shore erosion on three islands off the mouth of the Choptank River. He found that in 1848 Sharps Island consisted of 438 acres, but by 1910 was down to 53 acres, a loss of seven acres a year. Similarly, James Island was reduced from 976 acres in 1848 to 490 acres in 1910, a loss of eight acres annually, and Tilghman Island shrank from 2,015 acres in 1847 to 1,686 in 1900, a loss of six acres a year. Erosion rates such as these represent far more than a loss of real estate.

Throughout the bay, on mainland and island alike, coastal wetlands are threatened by erosion and rising sea levels. These wetlands—mainly marshes and the shallow waters at their edges—form a fragile ecosystem that supports the majority of the bay's living organisms. As the sea level rises inundating coastal wetlands, it destroys important habitats and sources of food for hundreds of species of fin and shellfish, crustaceans, birds, insects, and animals. When the rise is too rapid, which all too often is the case, it allows no time for the creatures that depend on the marshes or nearby shallows to adjust to the change or find new habitats. They go the way of the wetlands that supported them. Bit by bit, the bay's larger ecosystems are being altered and sometimes irrevocably changed.

The one-foot sea-level rise over the last century has already resulted in the loss not only of island marshes and natural shoreline habitats but also of a major portion of dry land. Evidence again is found in the bay's islands, hundreds of which have disappeared altogether either as a result of submergence or erosion. Man has also been affected by this process. Before European colonists arrived in the Chesapeake, Native Americans camped along the shores of the bay and its tributaries, but when a Western Shore bluff gave way to erosion or water submerged a favorite campsite on the Eastern Shore, they simply moved. English settlers were another matter and regularly built permanent structures close to the shore or on islands. St. Clement's, which was four hundred acres of thick woods when the first Marylanders landed, is now a nearly treeless forty-acre isle, the colonial fort and structures long gone. Sometimes colonists ignored or did not understand the impermanent nature of the land on which they built waterside structures. Families and entire communities eventually had to abandon their island homes.

Short of moving away from an eroding shoreline or preventing further development as Maryland has done with its "critical areas" regulations, people have made various attempts to deal with the rising waters and the loss of existing island homes. Wooden or concrete bulkheads, revetments of stone, and breakwaters offshore all have their advantages and disadvantages related either to cost or ecological impact. Where islands are concerned, owners and residents most

often yield to the inevitable and retreat to the mainland. As a result, the serious depopulation of the islands that began in the early decades of the twentieth century continues. Rather than remaining a place for permanent human habitation, some of the few islands that can be saved are being preserved as refuges for wildlife.

On September 18, 2003, while this material was going to press, the Chesapeake Bay area was hit by Hurricane Isabel, one of the worst in years. The cost of the damage in Maryland was estimated to be about $275 million and in Virginia to be more than a billion dollars. Damage was particularly high on the lower Eastern Shore. At the height of the storm tide, Watts, Tangier, and Smith Islands were completely submerged. Tangier watermen, worst hit on the southeastern side of the island, lost most of their over-water shanties, along with some 16,000 crab traps and other things stored within them. As important to their work as barns to farmers, these shanties, watermen learned, did not qualify for federal rebuilding assistance. "I lost diesel parts, tools, outboard motors," said the mayor of Tangier, Ed Parks. "Pretty much everything you'd need to make your living on this water was right there—everything you build up over time that you could sell whenever you retire. "On Fox Island the waves undermined the clubhouse of the Chesapeake Bay Foundation. Isabel spared the Hooper Island bridge, but most of the road was washed away. On the island itself, Chan Rippon's packing house lost a wall and part of its roof. Farther north, on Poplar Island, three dikes were washed away. St. Michaels was mostly under water. The store of the Maritime Museum had two feet of water over the floor.

On the lower Western Shore, many

bayside and riverside marinas and restaurants lost facilities. Gwynn's Island was hard hit, losing its restaurant and motel. An eleven-foot surge reached the upper Potomac River, causing Cobb Island to suffer severely. On the Patuxent River, Broomes Island was evacuated and severely damaged. Farther north, Annapolis received high water to the point that people paddled canoes and kayaks around the traffic circle on lower Main Street. Annapolis's Maritime Museum was flooded, and all waterside facilities were damaged by flood waters. On the upper Eastern Shore, Kent Island and the Chester River survived a storm surge of over seven feet, and Kent Island was completely isolated for a time, with both the Bay Bridge and the Kent Narrows Bridge closed. On the upper Western Shore extensive damage was reported all the way up to Havre de Grace, which lost the promenade surrounding the Decoy Museum.

Although many facilities were lost or badly mauled, waterside residents generally adopted an optimistic spirit and began rebuilding. Only a few businesses were forced to close permanently, and many have plans to expand or add improvements once rebuilt. "Isabel gave me a bang," seventy-five-year-old Robert Crockett of Tangier told *Baltimore Sun* reporter Chris Guy in April 2004. "For a while, I didn't even want to go out to the creek, I was just so blue. But this is one job you never want to leave as long as you can walk. I'm just hoping there's some crabs out there for us this year." The damage to the islands themselves was certainly extensive, and the final assessment of Isabel's impact awaits a complete survey of the area. Meantime, one can say with assurance that the future is open to change—both positive and negative. Incremental decisions of governments and individuals will tip the balance one way or the other. If there remain any doubts about where the bay is headed, the history of its vanishing islands will remove them.

A word on method. This book endeavors to document the losses to erosion that the islands of the Chesapeake have suffered over time. An abundance of antique charts with actual scale information available in various state and federal archives proved extremely valuable in this effort. For as many charts as I could locate for each island, I traced the area of the island using a planimeter (a cartographic tool that allows one precisely to measure areas on a flat projection) and converted to the scale of the chart. This approach supplied two to six data points for each island. No doubt many of the events leading to catastrophic loss of land are owed to storm conditions between surveys; we cannot know or accurately project annual rates of loss between surveys. Nevertheless, the method provides both a look at the dynamics of a given island over time and a comparison among islands in different parts of the bay.

# THE UPPER BAY
## FROM THE SUSQUEHANNA TO THE CHOPTANK

In 1608 Jamestown's Capt. John Smith explored the bay and probed its western tributaries hoping to find a way to China and the fabled riches of the Orient. Failing that, he made good use of his travels, describing the waters, land, native peoples, flora, and fauna he found. He and his companions were likely the first Englishmen to visit Garrett, Maids, Pooles, Cacaway, Wye, and Bruffs Islands, which, unlike the majority of bay islands, are much the same size as they were when Smith and other early settlers explored them for the first time. When he made his map of the bay in 1612, Smith named some of the islands. He named what is now Pooles Island after Nathaniel Powell, a member of his party of explorers. As so often occurred, misspellings by subsequent map makers and various officials resulted in the eventual name change.

Thanks to John Smith, we can picture the bay in the seventeenth century and gain a sense of the "delightsome" land he found. He wrote of the bay's bounty, noting that he and his companions had killed 148 ducks with just three shots as the waterfowl foraged for the wild celery and eel grass that grew in the vast shoals off the mouth of the Susquehanna River. These waters and the shores of islands and mainland continued to attract huge flocks of ducks and geese. An awestruck Maryland colonist wrote of seeing a flock of ducks a mile wide and seven miles long that blackened the skies over the Upper Bay. Around the turn of the eighteenth century, observers noted that 90 percent of the ducks that used the Atlantic flyway fed and rested in this area. Into the early 1900s, record keepers estimated that more than 800,000 ducks and geese made the annual stopover.

The profusion of waterfowl inevitably brought hunters, and by the late 1800s and early 1900s, they were bagging more than five thousand birds a day on the Susquehanna Flats alone. All over the bay, the same autumn scenario played out. Hundreds of sportsmen waited in onshore and offshore blinds or in rowboats camouflaged with reeds. They took a good many ducks and geese, but the much more efficient commercial hunters, using cannonlike "punt" guns, could kill up to one hundred birds with a single shot and wipe out an entire flock of resting birds in a single night. The sink-box gunner was another highly efficient hunter. Floating low and unseen in his coffinlike boat, he waited amid several hundred decoys for unsuspecting ducks or geese to fly in. At the opening of the 1879 season, one William Dobson opened up on a flock, firing as fast as he could reload his two double barreled shotguns until the barrel of one grew so

*Previous pages*: Bodine's charming view of Winchester Creek, 1953. (Bodine Collection, The Maryland Historical Society, Baltimore, Maryland)

hot that it burst. Dobson kept on firing, dipping his remaining gun in the water to cool it. His tally for that one day was more than five hundred ducks—the record for sink-box gunning.

The United States finally banned commercial hunting under the Migratory Bird Treaty Act of 1918, and soon thereafter a new breed of hunter—the outlaw gunner—appeared on the Susquehanna Flats and elsewhere on the Chesapeake. A market that remained strong, even during the Great Depression, made commercial hunting worth the risk for a number of bay farmers who needed a source of income to get them through the winters. Defying the government ban over the ensuing years, hunters, legitimate and outlaw, thinned the flocks to such an extent that Maryland's government established a bag limit of twenty-five ducks a day per hunter. Later the state cut the limit to twelve ducks, outlawed baiting, and in 1935 banned the sink-box. Today Maryland limits duck hunting to a few days in November and December with specific bag limits for each species.

Of greater importance than the fall advent of migrating waterfowl was the springtime arrival of shad on their way from the Atlantic to spawn in the upper reaches of the Susquehanna River. Many area families depended upon the income of husbands, fathers, and sons who worked the waters of the Susquehanna around Garrett Island and tiny man-made Fishing Battery Island below the Susquehanna's mouth.

For more than three hundred years, the Chesapeake was a busy north-south highway for vessels powered by wind, steam, and a variety of fuels. They followed the course of the drowned Susquehanna River from the Atlantic to the Upper Bay. Some vessels turned west toward the Severn River and Mary-

land's capital city of Annapolis, some toward the Patapsco and Baltimore, its major port. According to the *Martinet Atlas* of 1873, steamships continuing north could stop off to refuel at a coaling station on the northern end of Spesutie Island. They and other traffic then proceeded to Elkton at the top of the bay to offload passengers and cargo continuing by land. By the early 1800s, travelers could proceed on their northward journey by water, passing through the narrow locks of the Chesapeake and Delaware Canal.

The fur trade brought the earliest colonists to the Upper Bay. William Claiborne created a trading post and settlement on Kent Island, which he claimed as a part of Virginia. From that location, as well as Palmers Island in the Susquehanna River, he controlled most of the fur trade on the bay, and when the first Maryland colonists arrived in 1634, Claiborne refused to accept Lord Baltimore's authority. A bitter rivalry developed between his Kent Island settlement and the Maryland government in St. Mary's City. In 1635 the territorial feud erupted in a naval battle purported to have been the first in the Chesapeake Bay. The dispute lasted more than a dozen years, during which Claiborne's settlers waged war against Maryland forces ashore and afloat. In 1638 a contingent of Maryland's militia brought the rebellious Kent Islanders to heel, but the trouble continued between the political factions that Claiborne represented and those of Lord Baltimore. Shifts in political power in England resulted in a corresponding change in Maryland's government, and Lord Baltimore temporarily lost control of his colony. Claiborne was still very much a factor in Maryland's affairs. In 1652, when representatives of the province's new government met to sign a treaty with the

Susquehannocks, he was supported by parties on both sides in the negotiations. Affirming their regard for Claiborne, the Susquehannocks ceded all the land north of the Patuxent River to the Marylanders, "excepting the ile of Kent and Palmers Island which belong to Captaine Claiborne." Maryland's problems with Claiborne, like greater conflicts to follow, lasted many years on the bay, which was a bounteous resource and a busy and convenient highway connecting the middle colonies (and later, states) and the North and South.

This was no more true than in times of war, and many of the Chesapeake's islands figured in the movement of forces and supplies along the waterway. They harbored one faction or another during the Revolutionary War, War of 1812, and the Civil War. During the American Revolution, pro-British feeling was strong on the Eastern Shore, and loyalists raided bay shore and island properties of prominent patriots such as William Paca. A sailing barge belonging to Maryland's navy regularly cruised between Kent Point and lower Tilghman Island to protect the properties of Paca, Matthew Tilghman, and others.

Two of Maryland's leaders who owned islands in the Upper Bay were John Beale Bordley and William Paca, a Maryland signer of the Declaration of Independence. Bordley devoted his efforts to making gunpowder and provisioning the Continental Army, donating livestock and other foodstuffs raised on Pooles Island to help feed the American troops. He was part owner of Wye Island with William Paca and carried on similar provisioning operations there. Paca also helped supply the Continental Army with beef and wheat raised on Wye Island. Chesapeake islands produced more than provisions. Lambert Wicks, from

Eastern Neck Island, was a hero of the Continental Navy who successfully raided British shipping in the English Channel and safely delivered American emissary Benjamin Franklin to France.

Often during the War of 1812, the British fleet dominated the bay, moving freely and taking what provisions they needed whenever and wherever they dropped anchor. At one point a royal squadron paused off Spesutie Island and sent soldiers ashore to seize cattle and horses before proceeding to attack and burn Havre de Grace. Farther down the bay, British warships anchored off Poplar Island in the spring of 1813, and their foraging parties took large numbers of livestock belonging to the island's owner. In 1814 the British took advantage of the strategic location of Pooles Island and occupied it, confiscating livestock and produce to sustain the force manning a small gun battery set up to harass American shipping en route to the head of the bay and points north.

The Chesapeake both linked and divided North and South during the Civil War. Mirroring the national schism, opposing loyalties split friends and families. The war caused a rift between members of the Eastern Shore branch of the Paca family, one of whom still occupied Wye Island. William B. Paca of Wye Island stood firmly with the North, whereas Edward Paca, a near relative in blood and property, upheld the Southern cause from his Wye Plantation on the mainland. Many Marylanders were Southern sympathizers and readily sheltered rebel blockade runners who grew familiar with secluded coves, inlets, and islands that could hide their boats. A number sought the well-hidden harbor of Gibson Island, which, like other spots, returned to a more placid existence at the war's end.

Owners and other more or less permanent occupants found a variety of uses for islands of the Upper Bay, from the more usual enterprises of farming, working the water, or hunting to the 1840s black cat scheme involving Poplar Island and a later plan to hold a world championship boxing match on Pooles Island. Often their enterprises were as doomed as some of the islands themselves. As elsewhere in the bay, erosion took its toll, relentlessly eating away at the exposed edges of more fragile, soft-earthed islands such as Bodkin and Poplar. To slow or stop further erosion, the state of Maryland has taken over four of the endangered islands as depositories for dredge spoil. As a result, Hart, Miller, Poplar, and other islands elsewhere in the bay have become wildlife management and recreational areas.

Those islands of the Upper Bay that have been or are still occupied share much history and geography. Their individual stories illustrate the variety of natural and human forces at play upon them as well as upon the Chesapeake Bay, the surrounding land, and its people over the past four centuries.

## GARRETT ISLAND
### Chesapeake's Loftiest Isle

Thousands of travelers cross the Susquehanna River from Harford to Cecil County on four bridges. North of Garrett Island is the I-95 bridge, and downstream, Conrail uses the old Pennsylvania Railroad bridge. Two bridges span Garrett Island. One once served the Baltimore & Ohio Railroad and now the CSX; another serves US 40. Few who cross look down to see the rocky island that supports the bridges' massive piers of stone, steel, and concrete. Deep channels run along both sides of this rugged 190-acre river isle, flowing past shoals at its northern and southern tips and on to the river's mouth. Less than a mile long and not quite a half-mile wide, Garrett Island is dominated by a ninety-foot-high bluff, making it the loftiest island in the Chesapeake. Thanks to its deep-seated rocky foundations, Garrett Island has not gone the way of many other, often larger islands in the Chesapeake Bay watershed.

Those same rocks have been a hazard to sailors for centuries. Nearly four hundred years ago, Capt. John Smith com-

plained of losing his anchor among the rocks of the surrounding river bottom. Modern sailors are well advised to put a trip line on their anchors or suffer the same fate anchoring near the island today.

Garrett Island, which was granted to Virginian Edward Palmer in the early 1600s and for years bore his name, eventually became the site of the first English settlement within the present limits of Cecil County. Palmer had hoped to

Garrett Island (Palmer's Island) as it appeared on the 1799 Haudecoeur map, which captured its farm buildings and cultivated fields. Thanks to its sheltered location and rocky nature, Garrett—at least its shoreline—remained relatively unchanged at the turn of the twenty-first century. Highway and rail bridges now cross the island. (Courtesy of Anne Arundel County Public Library)

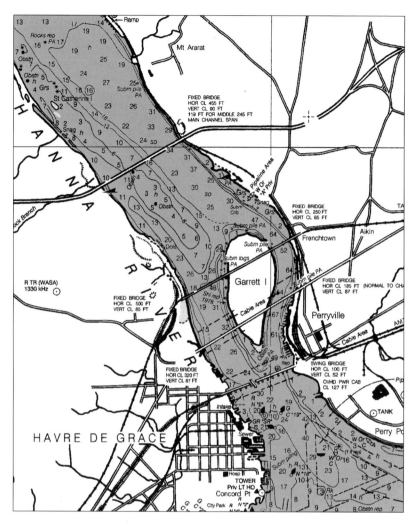

Garrett Island, 2002. (NOAA 12273, 2002. Courtesy of Maptech, Inc.)

found a center for learning on the island—a futile first attempt to establish a university in the New World. He never left England, however, and his heirs had no interest in his plans. The island remained a dominion of the Susquehannock Indians. A former indentured servant, George Alsop, described the Susquehannocks in a somewhat exaggerated account as "a people cast into the mould of a most large and Warlike deportment, the men being for the most part seven foot high . . . their voyce large and hollow, as ascending out of a Cave, their gate and behavior strait, stately, and majestic."

By the 1630s the Virginia fur trader William Claiborne had befriended these "majestick" people and profited greatly from fur trade with them. Recognizing that the English traders with their guns would be an effective buffer against enemies among the powerful Iroquois nation to the north, the "king of the Susquehannoes . . . did come with a great number of his Councellors and . . . did give Claiborne" the right to occupy Palmer's Island, plus "a greate deale of Land more." The king also "did cause his people to cleare some ground for . . . Claiborne to plant his corne upon."

Already in the fur-trading business on Kent Island, Claiborne established a trading post on Palmer's Island. He soon was in trouble with Maryland's governor Leonard Calvert, who sent an armed force to seize the island in 1638. Five years later, the Marylanders had to return to Palmer's Island to quell raiding Susquehannocks. They built a fort, the Fort Conquest mentioned in contemporary records. Trouble persisted, and in 1652 Maryland's government met to sign a treaty with the Susquehannocks. Claiborne was still powerful, and affirming their regard for him, the Susquehannocks ceded all the land north of the Patuxent River to the Marylanders, "excepting the ile of Kent and Palmers Island which belong to Captaine Claiborne." Lord Baltimore regained control of the island in 1658, the year that Henry Meese first surveyed it. By then, however, the fur trade was dying out. Powerful Senecas eventually drove the Susquehannocks from the area, and the colonists abandoned Palmer's Island.

Few signs of this early English occupation remain. In 1984 a systematic search for the site of Fort Conquest found no trace of it. Winter's ice working on the northern end of the island had removed any evidence of habitation there. Over the years, farmers clearing the land for planting and bulldozers engaged in construction of the present bridges finished the job of destroying all possible signs of early island settlements. Ultimately new development obliterated all evidence of the first settlement in the area.

The next stage of development on Palmer's Island began after the Revolutionary War when Isaac Watson established a farm that consisted of a house, barn, two acres of meadow, and fifty arable and seventy-one wooded acres. Owners continued to use Palmer's Island until the late 1800s. One of them was Mike Boyd, who, it is said, ferried his horses and farming equipment to the island daily.

Between 1820 and 1910 a fish-packing plant flourished on the island's northeast corner where seine fishermen hauled in shad from the Susquehanna River by the hundreds of barrels full. The fishing operation consisted of two "floats" moored north of the present CSX Railroad Bridge. These large log rafts supported living facilities and cleaning and salting tables. The fishermen landed their catch on an "apron" extending under water from each float. They fastened the seine net to one end of the float, and horses—later replaced by steam engines—pulled the net in at the opposite corner. The late Madison Mitchell, famous decoy maker from Havre de Grace, remembered being aboard Charles Silver's float when the fishermen put a net over one Monday morning and still had not emptied the net by the following Saturday afternoon. Since Charles Silver allowed no work on Sunday, they had to let Saturday's catch go or the fish would have smothered in the net over the weekend.

In 1885 the B&O Railroad began construction of a bridge and bought the island to support its piers, which were built from stone quarried on the south

side of the bluff. The railroad renamed its island in honor of B&O president John W. Garrett. In 1940 the state built a new toll bridge to carry the dual-lane highway US 40 over the island.

Today no evidence of the island's role in history is visible—no houses and no fishing sheds. There are only a few stone walls, a well, and several holes left by the occasional person looking for old bottles and other artifacts. A few explorers, attracted by stories of pirate treasure buried on the island, have been disappointed. By the 1960s the chief occupants of Garrett Island were deer. During the winter of 1961, observers counted eighteen deer that were near starvation on the snow-covered island. Concerned with their plight, B&O Railroad employees took up a collection, bought alfalfa bales, and dropped them to the deer from the railroad bridge.

The entire island was recently purchased by the U.S. Fish and Wildlife Department and has been designated as a wildlife refuge. The Department of Natural Resources and the state of Maryland will eventually own and manage it.

Coolahan and Hogan Fish Float near Garrett Island, about 1910. In spring, during the migratory runs of shad and herring, Baltimore and Philadelphia firms once hired hundreds of local men to work the nets. They unloaded tons of fish each day. (Historical Society of Harford County, Inc.)

REFERENCES

1799 C. P. Haudecoeur Map; 1858 Simon J. Martinet, Papenfuse/Coale; 1845 TC, USCGS, 189; 1846 HC, USCGS, 158; 1846 HC, USCGS, 326; 1899 TC, USCGS, 2382; 1975 PR Quad Havre de Grace, USCGS; 1976 NOAA C 4135; 1984 NOAA 12274 (229.57 acres); 1996 NOAA 12274 (189.86 acres). Erosion loss 0.4 acre per year.

Alsop, George, "A Character of the Province of Maryland, by George Alsop, 1666," in Clayton Colman Hall, ed., *Narratives of Early Maryland, 1633–1684* (New York: Charles Scribner's Sons, 1910; reprint, 1959). Carmer, Carl, *The Susquehanna* (New York: David McKay Co., 1955). Cronin, L. Eugene, letter to Sen. William S. James, 1972. Cronin, William B., "Garrett Island," *CBMag* (May 1985). "Eighteen Deer on Ice-Bound Isle Fed on Railmen's Largesse," *Sun,* February 18, 1961. Fausz, J. Frederick, "Following the Beaver's Path," St. Mary's College of Maryland alumni magazine *The Mulberry Tree* (1982). Fausz, J. Frederick, "Profits, Pelts, and Power: English Culture in the Early Chesapeake, 1620–1652," *Maryland Historian* 14 (1983): 15–30. Jay, Peter, "A Trespasser Alone with the Deer and the Litter," *Sun,* March 23, 1997. Johnson, George, *History of Cecil County* (Elkton, Md.: private printing, n.d.; reprint, Baltimore: Regional Publishing Co., 1967). Klinglehofer, Eric, "The Search for 'ffort Conquest' and the Claibourne Virginia Settlement: An Archeological Study of Garrett Island" (Ph.D. diss., Johns Hopkins University, 1984). Mason, Samuel, Jr., *Historical Sketches of Harford County* (Lancaster, Pa.: Intelligencer Printing Co., 1940). Miller, Alice, *Cecil County* (Port Deposit, Md.: Port Deposit Heritage, 1949). Mitchell, R. Madison, interview by the author, January 1985. Semmes, Raphael, *Captains and Mariners of Early Maryland* (New York: Arno Press, 1937; reprint, Baltimore: Johns Hopkins University Press, 1979). Shomette, Donald G., *Ghost Fleet of Mallows Bay and Other Tales of the Lost Chesapeake* (Centreville, Md.: Tidewater, 1996). See also William B. Marye, "'Patowmek Above Ye Inhabitants': A Commentary on the Subject of an Old Map," pt. 3, *MdHM* 32 (1937):299 n.; Cyprian Thorogood, *A Relation of a Voyage Made by Mr. Cyprian Thorogood to the Head of the Baye, 1634,* trans. Clifford Lewis III. Wright, C. Milton, *Our Harford Heritage* (Glen Burnie, Md.: French-Bray Printing Co., 1967).

## MAIDS ISLAND
*Swan Creek Island Linked to Logging Industry*

Tiny one-half-acre Maids Island lies off Cedar Point in Swan Creek, just below the mouth of the Susquehanna River. Because of its location, the story of Maids Island is actually part of a bigger one that ties it inextricably to the creek and the nineteenth-century mill town of Oakington.

Originally, the island was part of an 800-acre grant belonging to Col. Nathaniel Utie of nearby Spesutie Island. Utie obtained the patent for the land in 1660 and called it Oakington. A year later, he sold Oakington to Rutten Garrett. With Garrett's death, Oakington and the island went to his widow, Mary, who subsequently married Edward Beedle, and, after his death, George Utie. In 1696, when George Utie died, Mary and her brother sold Oakington to the first of several owners. Eventually it came into the hands of John W. Stump, who built a large stone house on the property in 1810. After the land passed through several more hands, a Henry James bought it in 1867. This deed includes the first known mention of Maids Island, described as "about 1/2 acre in size, more or less." The island's name apparently came from the Susquehannock Indians. As the story goes, according to tribal custom, older women took their tribe's young unmarried maidens to the island to train them to be good wives.

The Susquehannocks were long gone by 1808 when the Susquehanna Canal opened an unnavigable stretch of the river as far as the Pennsylvania line. Soon huge lumber rafts were a common sight on the waterway. Some of these arks, or barges, were as long as 180 feet and were made of timbers spiked together. They carried flour, wheat, pork, iron, slate, coal, and other products from Pennsylvania to ships waiting in Port Deposit.

Once unloaded, the arks were broken up and sold for lumber, much of which was subsequently floated down the river, through Havre de Grace to Swan Creek and three lumber mills at Oakington. The largest of the three was owned by the aforementioned Henry James. The mills brought workers, and soon a small community had developed, complete with a schoolhouse, doctor, and store. Once the logs reached the mill site on Swan Creek, they were corralled by large floating booms and pilings. It was said that there were so many that people could walk across the creek on them. As many as 1,000 log rafts came down the Susquehanna before railroads replaced river transport in the 1870s. When the canal closed, the Oakington mills also closed, as their chief source of lumber disappeared. Pilings are all that remain of the milling operation, and they still appear on modern charts as a danger to navigation. The town went the way of the mills, and in 1882 a John M. Michael bought about 350 acres of the Oakington tract for growing corn. Eventually he tore down the remaining buildings.

Oakington as a tract of land remained more or less intact, and by the twentieth century it was the estate of J. Millard Tydings, Maryland's U.S. senator from

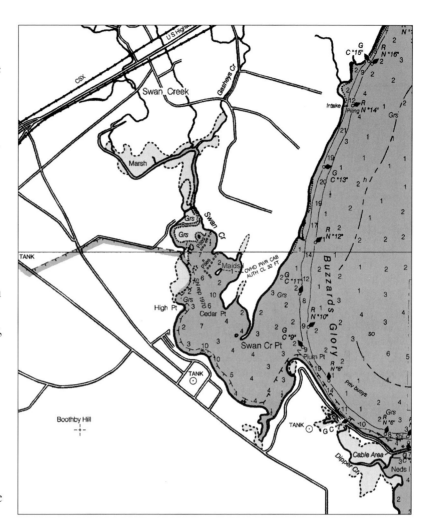

1927 to 1951, and after him, of his son Joseph, also a U.S. senator, serving from 1964 to 1970. The house on the island has its own power line to the mainland and is a private residence.

Maids Island, 2002. (NOAA 12273, 2002. Courtesy of Maptech, Inc.)

REFERENCES

1845 TC, USCGS 188; 1846 HC, USCGS 185; 1899 HC, USCGS 2 432; 1938 HC, USCGS 6365; 1996 NOAA 12274.

Arnett et al. Davis, Jack, interview by the author, January 24, 1998. *History of Oakington* (Havre de Grace, Md.: Harford County Historical Society, n.d.). "Oakington: The Town That Disappeared," *Aegis* (Harford County, Maryland), August 2, 1992. Papenfuse et al. *Plan of the Town of Oakington,* surveyed and drawn by J. S. Newlin, April 30, 1867.

## FISHING BATTERY
*Man-made Island Born of the Bounty of the Bay*

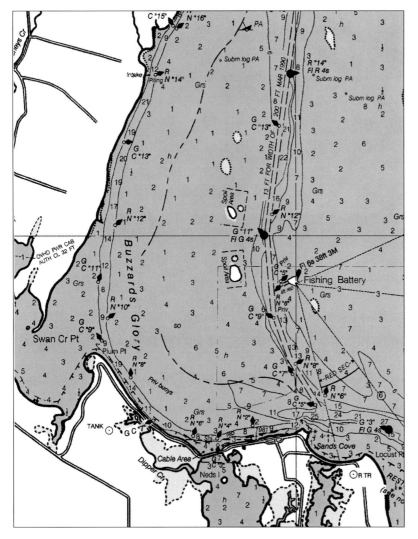

Fishing Battery Island,
2002. (NOAA 12274, 2002.
Courtesy of Maptech, Inc.)

Tiny Fishing Battery Island lies two and a half miles south of Havre de Grace, just east of the channel dug eons ago by the once mighty Susquehanna River. In the late eighteenth and early nineteenth centuries, fishermen of the Upper Bay worked the waters of this area, netting tons of shad on their annual spring run from the Atlantic up the Susquehanna. The great runs of spawning shad extended as far as Juniata, Pennsylvania, and even into New York state. To harvest the fish in the Susquehanna alone, fishermen used about 3,000 nets of various kinds. Working the Upper Bay, fish-

ermen used at least twenty large seine nets, drawing them to the shore or to large log rafts known as floats. There, the seine haulers anchored their nets, emptied them, and processed their catch.

To take full advantage of the seemingly endless runs of shad, entrepreneurs Robert Gale and John Donahoo built an island below the mouth of the Susquehanna in the early 1800s. A strictly utilitarian construction known as a fishing battery, the island provided a hauling ramp, housing for workers, and facilities for cutting, salting, and packing fish. Surveyed in 1836, Donahoo's Battery, as it was first known, contained one acre and eighteen square perches (a square perch being equal to a square rod, or 30.25 square yards). Fishing Battery Island has changed very little in size since 1836 and is one of the smallest islands in the bay. People once called it Edmondson's Island, presumably after an early owner, but usually they have used the more descriptive names, Shad Battery, Fishing Battery, or simply Battery Island.

The shad industry reached its zenith between 1888 and 1909, but the construction of dams, especially the Conowingo Dam built in the 1920s, caused the loss of shad breeding grounds and subsequently reduced the number that reached the upper Susquehanna to spawn. Their numbers diminished to critical levels, and in 1980, Maryland finally prohibited shad fishery, hoping to revive the population. Rockfish suffered the same fate, and in 1985 the state banned rock fishing as well. Although recent counts of young shad and rockfish are up, their numbers will probably never reach the phenomenal heights

of the nineteenth century. Nonetheless, the rockfish recovery is now considered a triumph of fisheries restoration on the Chesapeake Bay.

Fishing Battery Island has been more than a platform for fishing operations. In 1852, Maryland's Lighthouse Board recommended the purchase of a forty-five-by-forty-five-foot section of Donahoo's Battery for ten dollars. A Havre de Grace city official and a contractor of sorts, John Donahoo had been involved in building several bay lighthouses. His bid won the contract to build a lighthouse on Battery Island in 1853. Characteristic of lighthouses built during this period, the beacon topped a thirty-two-foot-high tower rising from the roof of the keeper's house. Donahoo's lighthouse served until 1921, when the state replaced the old house and lantern with

an automated beacon on a thirty-eight-foot steel tower. Today, the beacon is a fixed white light.

Maryland became more heavily involved with the island in 1861 when the Bureau of Fisheries leased it for use as a shad hatchery and for related fish culture studies. In 1891, the bureau finally purchased the island for $1,500. However, the story of Fishing Battery Island does not end with lighthouses and fish. In 1942, the Department of the Interior took over Fishing Battery Island. Now part of the Susquehanna Wildlife Refuge, it is not open to the public. In recent years, the Army Corps of Engineers created two small spoil islands to the west and northwest of Fishing Battery Island. Formed during the dredging of the approach channel to Havre de Grace, the two elliptically shaped islands

Aerial view of Fishing Battery Island, 1999. The lighthouse on the island, built in 1853, eventually fell into disuse and was abandoned. An automated modern tower built in 1915 stands alongside the ruins. (Author photograph)

are overgrown by weeds, bushes, and small trees. Local boaters use them for picnicking.

In the 1990s, a group of citizens of Havre de Grace formed the Battery Island Preservation Society principally to preserve the lighthouse. Governmental red tape, difficulties in getting to the island, and a lack of funding for restoration finally put an end to their efforts, and without official supervision, the island has fallen victim to vandalism.

REFERENCES

1836 Gale/Donahoo survey (1+ acres); 1845 TC, USCGS, 188; 1846 HC, USCGS, 185; 1867 HC, USCGS, 198; 1898 TC, USCGS, 2384; 1939 HC, USCGS, 6368; 1943 TC, USCGS, 8288; 1976 NOAA C4136 (shows spoil islands); 1984 NOAA 12274; 1996 NOAA 12274.

Cronin, L. Eugene, "The Bounty of the Bay, 1584–2034," paper presented at "Maryland Our Maryland" seminar series, College of Notre Dame of Maryland, December 1984. Hornberger/Turbeyville. Mitchell, R. Madison, "I Remember . . . Flocks That Blotted Out the Sun," *Sun*, Feb. 25, 1962. Mitchell, R. Madison, interview by the author, January 1985. Wright, C. Milton, *Our Harford Heritage* (Glen Burnie, Md.: French-Bray Printing Co., 1967).

## SPESUTIE ISLAND
*From Indian Hunting Ground to Rocket-testing Range*

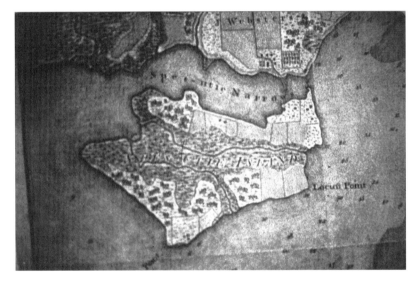

Detail from 1799 Haudecoeur map, showing Spesutie Island and its several farms, cultivated fields, and orchards. Col. William Smith and his family lived there for more than a century in a house that after 1870 served as a hunting lodge. In 1918 local firefighters deliberately burned it in a drill. (Courtesy of Anne Arundel County Public Library)

Spesutie Island, pronounced locally as Spes-sú-sie, is a 1,500-acre island in the upper Chesapeake Bay. Indians and early English settlers preceded farmers, hunters, fishermen, and finally the U.S. Army in shaping the island's long and varied history. Part of the Aberdeen Proving Ground today, Spesutie Island was where Susquehannock Indians once hunted bear and other game. Modern-day explorers have turned up arrowheads and other ar-

tifacts, but the Susquehannocks left no trace of permanent settlement.

The island received its present name in 1658 when Lord Baltimore granted it to Nathaniel Utie. A native Virginian who emigrated to Maryland in 1656, Utie named his 2,300-acre grant Spesutie Manor. Utie combined his name with *spes,* the Latin word for hope, to arrive at a name for the manor house he built on the northern end of the island near the present causeway.

Merchant, Indian trader, and planter, Nathaniel Utie was a man of wealth and stature in early Maryland. As master of Spesutie Manor, Utie was the largest landholder in the newly created Baltimore County. His position earned him a seat in the General Assembly and, subsequently, on the Governor's Council for his ability and "affectionate service in the Assembly." Utie also attained the rank of colonel in the provincial militia as commander of the northern frontier.

At about the same time that Utie was establishing himself on the upper Chesapeake Bay, Dutch settlers were moving into areas between the Chesapeake and Delaware Bays. The government and citizens of the Dutch colony of New Netherlands (now New York) ignored the fact that they were encroaching on land encompassed as a part of Maryland in Lord Baltimore's original grant. Finally, in 1659, Maryland's governor Josias Fendall sent Nathaniel Utie to demand that the Dutch colonists accept Lord Baltimore's authority or leave his province. Arriving at the town of New Amstel (now New Castle), Delaware, Utie demanded in a "boisterous and stormy manner" that the Dutch either submit to Lord Baltimore or abandon the town. When the Marylander threatened bloodshed if the citizens of New Amstel did not comply, Peter Stuyvesant, governor of New Netherlands, sent troops from New Amsterdam to seize Utie and protect the Dutch settlements in the Delaware Bay area. Utie evaded Stuyvesant's troops and returned to Maryland and his duties as a council member.

There he was soon embroiled in more controversy, siding with Governor Fendall and the lower house of the assembly in claiming that Marylanders should control their own affairs rather than answer to the Catholic lord proprietor. In a bloodless rebellion in 1659, the Fendall faction took over the government and established a commonwealth. A year later, still without bloodshed, Cecil Calvert regained control of his colony's government. Though threatened with banishment and confiscation of their property, Fendall, Utie, and other leaders of the short-lived commonwealth ultimately suffered only the loss of their right to vote and hold office. In 1661, however, Calvert pardoned Utie, who again returned to the assembly as Baltimore County's representative.

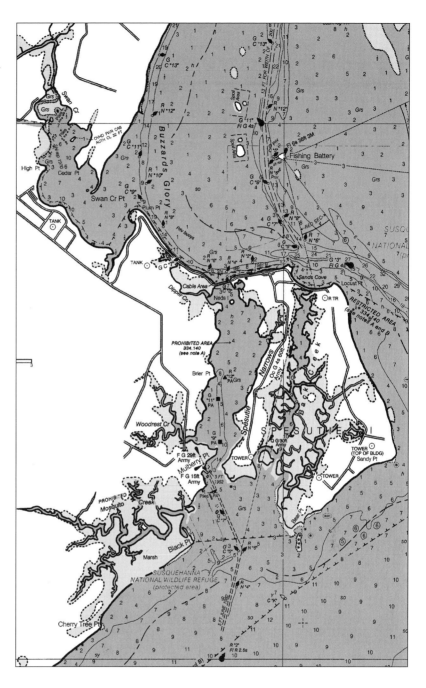

Nathaniel Utie's first wife, Mary, was stabbed to death by a slave. His second wife, Elizabeth, survived him and their only son who died in childhood. The manor passed to his nephew, George, and remained in the family until 1779 when Col. Samuel Hughes, the iron master of Catoctin Furnace, bought the 2,154-acre estate for 7,500 pounds sterling. In 1802, Hughes sold the island to Col. William Smith, and the Smith

Spesutie Island, connected by a bridge to the mainland (though not open to the public) and nearly cut in two by deepening marshland, 2002. Utie's original grant in 1658 was for a tract of 2,300 acres. In 1996 acreage stood at 1,502, marking a loss of 798 acres in 338 years or an average of 2.4 acres annually. (NOAA 12274, 2002. Courtesy of Maptech, Inc.)

family lived there for more than a century. Over the years, they enlarged Utie's manor house. They also suffered some uneasy moments during the War of 1812, when Admiral Cockburn's British squadron paused offshore. Fortunately for the island's inhabitants, the British did not know that William Smith was the brother of Maj. Gen. Samuel Smith, commander of the third division of the Maryland Volunteer Militia charged with the defense of Baltimore. After sending soldiers ashore to seize cattle and horses, the British left the Smiths unmolested and proceeded north where they attacked and burned Havre de Grace.

Upon William Smith's death, Robert Smith used the Spesutie Island house as a hunting lodge. John D. Smith, author of several famous books on the flora of South America, was the fourth generation of the family to own the island. These latter Smiths and several of their visitors were the first on record to encounter the Spesutie Manor ghost. They vowed they saw a white dog roaming the house at night, especially on the stairs. Throughout their ownership, the Smiths rented out land to tenant farmers and ran a small canning house to process the corn and tomatoes grown on the island. The family finally sold the island in 1900, and in 1924 it came into the hands of the Bartlet Ford family.

The late Wilson Ford, Bartlet Ford's son, spoke about life in the old Utie house, which he said was on Bear Point. In those days, baiting to attract ducks and geese was still legal. The Fords raised corn and wheat. In the fall, they loaded it into a rowboat and spread it around with shovels to bring ducks to where hunters waited in their blinds. The island had no electricity, telephones, or cars, and to go to school on the mainland, island children set out before day-

light with a horse and buggy. They left some feed with the tethered horse and crossed Spesutie Narrows by rowboat. On the other side, they waited to be picked up by a school bus. To cross the frozen Narrows in the wintertime, the children pushed themselves across the ice in a rowboat equipped with skids. Thus, if the ice broke, they were not dumped into the freezing water. In severe winters, the children stayed with relatives on the mainland. The spot where the children crossed the Narrows was also the site of one of the earliest ferries in Maryland. Using a rope and pulley system, the ferry operator hauled the large, flat scow capable of carrying a horse and wagon between shores.

In 1917 World War I radically changed the character of the surrounding land, 35,000 acres of which had been bought by the federal government to establish the Aberdeen Proving Ground. The purchase meant the loss of some of the finest corn-growing land in the East and a large number of corn and tomato canneries that had contributed greatly to the economy of the area.

The sale did not include Spesutie Island, which remained in private hands until 1927, when the Fords sold it to the Susquehanna Development Company. It, in turn, sold the island's six farms to their tenants or to new owners. J. Pierpont Morgan built a large frame house on the island in 1936, and he and his friends used it as a hunting lodge. The following year, the Spesutie Island Development Company, representing the Spesutie Island Hunting and Fishing Club, bought the island. The club maintained the farms as a source of corn to use as bait for ducks and geese.

Finally, in 1945, Stephen Burch Jr., heir of one of the hunting-lodge owners, sold Spesutie Island to the U.S. government. The price of the 2,065.9-acre is-

A Model T makes its way to Spesutie on a ferry not long before World War I. In 1945 the federal government condemned the island and made it a part of the army's Aberdeen Proving Ground. (Historical Society of Harford County, Inc.)

land was $95,000. It is now part of Aberdeen Proving Ground and contains a large air-to-ground rocket range and a fuse-testing and bomb-disposal area. Eventually the government built a causeway joining the mainland and the northern end of the island. Though the roadway improved access, the resulting obstruction interrupted the natural flushing effect of the waters flowing through Spesutie Narrows and caused a decline in their productivity.

During a fire-fighting exercise, trainers set fire to Utie's original manor house. Firefighters got the blaze under control, but all that remained was a charred shell. A large and significant tulip poplar and other trees surrounding the house also were destroyed. Eventually the roof of the house fell in, and now not a trace remains.

The island is now nearly the same size it was when the first settlers arrived. Protected by surrounding land masses, Spesutie Island has suffered little of the storm erosion that has reduced other is-

lands. It has lost only 709.94 acres in more than three hundred years, which, at a rate of about 4.79 acres per year, is very low for an island in the Chesapeake.

Despite the hunting on the island and explosions from the nearby proving ground, a large concentration of bald eagles chose Spesutie as a nesting spot. Before the Great Depression, eagle eggs were bringing up to ten dollars each, and would-be collectors kept track of large bald eagle populations. In 1930 a count showed that the island averaged a pair of eagles to every three miles of shoreline. The largest nest reported was in a white oak tree and was 115 feet above the ground. Hard economic times apparently put an end to the demand for egg collecting, but soon the insecticide DDT, which softened the shells of the bald eagle's eggs, posed a more serious threat to the birds' survival. Only after the government banned DDT and added bald eagles to its endangered species list did their numbers begin to increase. Today many eagles nest on and around Spesutie Island, ignoring the thunder of bomb disposal and rocket tests that rattle windows across the bay. Eagles and other wildlife may have free access to Spesutie Island, but the federally owned property is closed to the public except by special permission.

REFERENCES

1658 Utie grant (2,300 acres); 1779 Wilmer Barnes plat (2,154 acres) private collection; 1799 Haudecoeur Map, Papenfuse/Coale; 1846 HC, USCGS, 185; 1846 HC, USCGS, 186; 1848 Spesutie Q, USGS (2,212.8); 1848 Martinet Map, Papenfuse/Coale; 1898 HC, USCGS, 2393; 1938 HC, USCGS, 6365; 1942 Spesutie Q, USGS (2,065.9 acres); 1964 US Chart 572-SC (1,904.1 acres); 1975 Spesutie Q, USGS; 1984 NOAA 12274 (1,594.25 acres); 1996 NOAA 12274 (1,502.86 acres). Erosion loss 2.36 acres per year.

Arnett et al. Byron, Gilbert, *The War of 1812 on the Chesapeake Bay* (Baltimore: MdHS, 1964). Clark, Marjorie, interview by the author, March 1985. Earle, Swepson, *The Chesapeake Bay Country* (Baltimore: Remington-Putnam, 1936). Ford, Wilson, interview by the author, January 1985. Koeth, James, "A Ghost, Classified 'Secret' by the Army, Roams Spesutie Island," *Sun*, November 13, 1958. Martinet, S. J., H. F. Wallina, and O. W. Gray, *Topographical Atlas of Maryland, Counties of Baltimore & Harford* (1873). Mayre, W. B., "The Place Names of Baltimore & Harford Counties," *MdHM* 25 (December 1936). Mayre, W. B., "Former Indian Sites in Maryland as Located by Early Colonial Records," *American Antiquity* (1938, 1940–46). Mayre, W. B., "Shell Heaps on the Chesapeake Bay," *American Antiquity* (1939). Meanly, Brooks, *Birds and Marshes of the Chesapeake Bay Country* (Centreville, Md.: Tidewater, 1975). Morehead, W. K., ed., *A Report on the Susquehanna Expedition* (Andover, Mass.: n.p., 1938). Papenfuse/Coale. Papenfuse et al. Scarborough, Katherine, "White Ghost Dog Roams the Manor," *Sun*, November 28, 1954. Storck, Nan Davidson, interview by the author, July 18, 1987. Wilke, Steve, and Gail Thompson, *Prehistoric Archeological Resources in the Maryland Coastal Zone* (MdDNR, 1977). Wilstach, Paul, *Tidewater Maryland* (Cambridge, Md.: Tidewater, 1969). Wright, C. Milton, *Our Harford Heritage* (Glen Burnie, Md.: French-Bray Printing Co., 1967).

## POOLES ISLAND

*Site of Colonial Farm, Historic Lighthouse, and Would-be Championship Boxing Match*

Pooles Island has been a no-man's-land since 1917, when it became part of the U.S. Army's Aberdeen Proving Ground. Approximately one mile offshore, the island marks the southeast corner of the vast restricted land and water area occupied by the military installation. Access to the island and the waters around it is governed by the army's firing schedule and requires special permission—and with good reason. Aside from the danger of coming under cannon fire, anyone venturing ashore on Pooles Island walks into a dangerously deceptive woodland liberally seeded during decades of munitions testing with unexploded bombs and other ordnance.

Tides around the island typically rise over a period of about six hours. The ebbing tide takes an hour longer, receding little more than a foot. Though the rise and fall are not great, the tidal current is strong east of the island where the bottom drops suddenly from an average of eight feet to a thirty-to-fifty-foot depth. At full ebb, the current approaches one and one-half knots. Centuries ago, oysters were a staple in the diet of Indians who occupied the island and who left many oyster shell middens scattered along its shores. Near

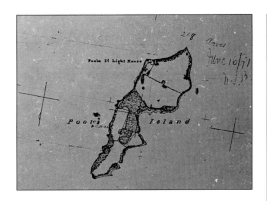

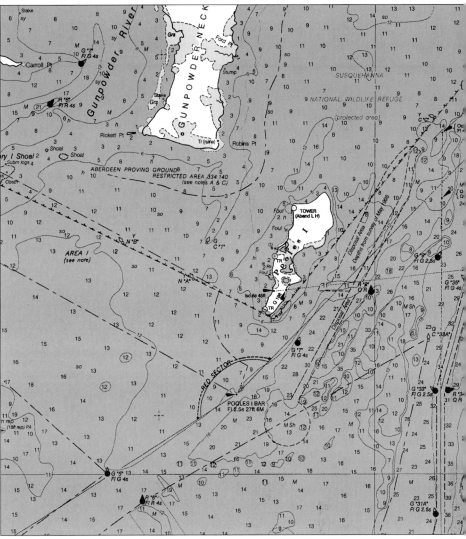

Pooles Island are large oyster shell "humps"—reminders that oysters were once abundant in the upper Chesapeake, when its waters furnished the salinity oysters need to survive. In recent years, the state of Maryland has been dredging these old oyster beds near the island to plant the shells in areas more suitable for setting spat, as the young oysters are called. Under tow, the barges—usually strung together three at a time—make the trip down the bay accompanied by great flocks of seagulls riding atop the heaps of shells.

Until army occupation, the island was a relatively busy place. Early in the 1600s, Capt. John Smith and his companions were perhaps the first Englishmen to visit Pooles Island. Smith gave it the name "Powels Island," after Nathaniel Powell, one of his fellow explorers, and included it when he made his map of the bay in 1612. Maryland adventurer and fur trader Cyprian Thorogood called it Pooles Island in his account of his voyage up the bay in 1634, and Nathaniel Powell was soon forgotten. A surveyor, who measured an area of 200 acres in 1659, called it "Pools Island," and the name became more or less official when Maryland's first cartographer, Augustine Herrman, included Pooles Island on his 1670 map of Maryland. Subsequent map makers spelled it variously as Pools and Pooles.

John Beale Bordley, one of colonial Maryland's most prominent citizens, bought Pooles Island in 1771. An attorney, legislator, council member, and judge, Bordley was also keenly interested in mathematics, philosophy, and the natural sciences. He owned land in several counties where he experimented with agricultural methods and breeding livestock. When Bordley purchased Pooles Island, his survey showed that it consisted of 295 acres, much of which was planted in tobacco and wheat. A previous owner had introduced wheat about 1751. (For many years before that, owners grew Indian corn.) Tobacco culture ended with Bordley's purchase; he had other plans.

*Above left:* Pooles Island, 1854, when an orchard yielded prize peaches for the Baltimore market. (RG 23, Chart T450, NARA)

*Above right:* Pooles Island, 2002. Like Spesutie under federal control and off-limits to the public, the island flourishes as a haven for wildlife. In 1771 Pooles consisted of 295 acres; by 1996 that figure had dropped to 213. (NOAA 12278, 2002. Courtesy of Maptech, Inc.)

The Pooles Island lighthouse (built in 1895, abandoned in 1939, exterior restored in 1999). It holds the distinction of being the oldest lighthouse in Maryland. (Author photograph, 2000)

lished a gun battery overlooking the nearby busy trade route and plundered livestock and produce from island farms.

The same advantageous location led to the construction of a lighthouse on the island's northwest corner in 1825. John Donahoo, who built several bay lighthouses under contract to the federal Treasury Department's Lighthouse Board, constructed the beacon for about $5,500. The board provided for a lighthouse keeper until 1917 when it decided to automate the light. Finally, in 1939, the federal government permanently abandoned the Pooles Island light and ordered the lens removed.

In 1848, pugilists Tom Hyer and Yankee Sullivan were ready to square off for the world heavyweight championship. At that time boxing was illegal, and if the fight was to come off, they needed a remote site beyond the reach of the law. What better place than a deserted island? Except for a lighthouse keeper, assorted waterfowl, and other wildlife, Pooles Island filled the bill. The scene was set for events that rivaled the antics in a Keystone Cops film. Fight promoters, trainers and sparring partners, fighters, and a small corps of newspaper reporters arrived aboard four vessels. While the fight promoters were getting everything ready, the fighters and their attendants settled into a pair of abandoned buildings, the only two on the island.

The excitement began as a government steamer arrived with an outpouring of state police and militia. They swept down on the buildings, but word of their arrival preceded them. Hyer had slipped out a back window and escaped in a waiting boat. The police had to be content with arresting his trainer. In the second house, they found the other fighter and his sparring partner, but Sullivan outsmarted them. As the police approached, he yelled to his companion,

Deer and wild turkeys abounded on the island, and, perhaps with the idea of creating a hunting preserve, Bordley added hares and partridges imported from England. He also kept a herd of cattle. During the Revolutionary War, he used the island for gunpowder production and raising livestock and other foodstuffs to help feed the Continental Army.

Overlooking a relatively narrow stretch of the Chesapeake, Pooles Island attracted the attention of the British during the War of 1812. In 1814 they estab-

"Run, Sullivan!" The man took off with the police in hot pursuit, and Sullivan walked away. Undeterred by police interference, the fight promoters simply moved their operation to a farm on the Eastern Shore. Hyer won the championship and a purse of $10,000, of which state authorities took a $1,000 fine when they eventually caught up with him.

Agricultural interests put Pooles Island to more productive use. In 1873 a small freight boat unloaded seven thousand peach tree saplings and the field hands to plant them on the northern end of the island. When the trees reached maturity, the peaches they bore were shipped to Baltimore, where their flavor commanded a premium price.

Since the army's occupation, nature has been allowed to take its course. The upper and lower ends of the island are wooded and in between are several freshwater ponds surrounded by swamp. Although the army has taken no action to prevent erosion, Pooles Island has not suffered as much land loss as have other Chesapeake islands. Between 1845, when the island measured about 277 acres, and a 1996 survey showing approximately 213 acres, the loss has been 0.4 acres per year.

Army ordnance testing may keep human beings away, but it does not seem to have driven off the wildlife. Protected by the army's guns, the island and much of the rest of the proving ground are part of the Susquehanna National Wildlife Refuge. Early morning boaters moored in safe waters south of the island may see deer on the shore or swimming between it and the mainland. From February to early summer, with little or no disturbance from people, large numbers of great blue herons occupy their rookeries high in the trees on the southern end of Pooles Island. Bald eagles also nest on the island, which is a haven for many native and migrating songbirds as well as foxes and muskrats.

The lighthouse still stands as a picturesque ruin, which the federal government has added to the National Register of Historic Places. The proving ground's public works department has repaired it, and a volunteer contractor gave it a fresh coat of paint. Thus far, preservation efforts have not extended to a badly deteriorating World War I observation tower built to catch trespassers and pinpoint the locations of exploding ordnance. Today the island is strictly off-limits, but those interested in visiting the lighthouse may obtain special permission from proving ground officials.

REFERENCES

1659 survey (200 acres); 1771 Bordley survey (295 acres); 1845 HC, USCGS, 166; 1845 TC, USCGS, 187 (276.83 acres); 1846 HC, USCGS, 187 (261.58); 1854 TC, USCGS, 450 (67.28 acres); 1878 Martinet Map, Papenfuse/Coale; 1897 TC, USCGS, 2296 (240.56 acres); 1898 HC, USCGS, 2399; 1938 HC, USCGS, 6373 (228.63); 1972 DNR-H, A1-4RL-20; 1976 NOAA 1278; 1984 NOAA 122742 (16.67 acres); 1996 NOAA 122782 (12.92 acres). Erosion loss 0.404 acres per year.
"Atlantic Coast: Sandy Hook to Cape Henry," *US Coast Pilot* 3, NOAA (n.d.). Cronin, William B., "Volumetric, Areal, and Tidal Statistics of the Chesapeake Bay Estuary and Its Tributaries," *Special Report* 20, Chesapeake Bay Institute (1971). *Guide for Cruising Maryland Waters*, MdDNR, annually. Hayes, Anne H., and Harriet R. Hazleton, *Chesapeake Kaleidoscope* (Cambridge, Md.: Tidewater, 1975). Hornberger/Turbeyville. Mayre, William B., "Place Names in Baltimore and Harford Counties," *MdHM* 25 (December 1936). McGrath, Francis S., *Pillars of Maryland* (Richmond: Dietz Press, 1950). Papenfuse/Coale. Papenfuse et al. Usilton, Fred G., *History of Kent County* (Chestertown, Md.: *Kent News,* 1980). Wright, C. Milton, *Our Harford Heritage* (Glen Burnie, Md.: French-Bray Printing Co., 1967).

## HART AND MILLER ISLANDS
*Eroding Islands Saved by Dredged Spoil*

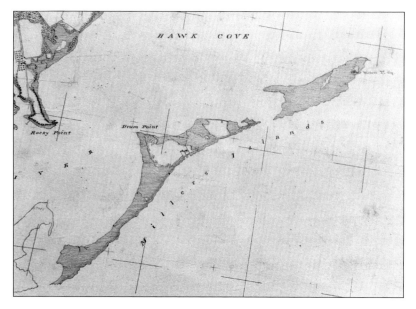

*Top:* Millers Island, 1854, when acreage totaled about 335. (Courtesy of National Archives (NACP) RG 23, T450)

*Bottom:* By 1898, charts showed two distinct islands, Millers, on the north, about as it had appeared half a century earlier, and Hart, to the south, as a mere remnant. Another piece of lower Millers Island came to be known as Pleasure Island. (Courtesy of National Archives (NACP) RG 23, T2308)

At the mouth of Back River, Hart and Miller Islands were once part of a long peninsula extending to the northeast from what is now North Point State Park. Since the late 1600s, however, continual erosion created and then steadily reduced the resulting islands. An 1846 survey shows Hart to be 197 acres and Millers, or Miller, as it was later called, 138 acres. By 1984 they had eroded to 107 and 30 acres, respectively.

The islands were settled in the early 1800s by Joseph Hart, an eccentric Eng-

lish merchant who once owned a tavern in nearby Baltimore. Local legend tells of his mistrust of banks, which led him to bury his life savings of $15,000 somewhere on the island. As the story goes, he also found a keg of gold pieces and hid that booty, too. Despite the efforts of would-be treasure hunters, Hart's legendary fortune has never been found.

A farm that operated on Hart Island in the last half of the nineteenth century consisted of three houses and several fields and orchards. No one seems to have lived on Millers, which was shown on early charts as mostly swampland. Over the ensuing years, the islands had a number of owners, most illustrious of whom was George P. Mahoney. As chairman of the Maryland Racing Commission, he was known for cleaning up horseracing in the 1940s. Mahoney later ran for governor, losing to Spiro T. Agnew. A principal in the Mahoney Brothers Engineering and Construction Company, he also had great plans for Hart Island. Like his race for the governorship, however, his plan to build a resort and summer homes on the island fell short of the mark.

The Maryland Department of Natural Resources bought the islands in 1975 and constructed a large containment area of over a thousand acres. In 1983 Hart and Miller began receiving spoil from the dredging of the Baltimore channels, which eventually filled the northern two-thirds of the containment area. In the 1990s additional bulkheading marked the start of work on filling the last portion. Eventually a total of 27 million cubic yards of dredged material will become a 1,140-acre wildlife sanctuary and recreation area. Accessible only by boat, it will be an extension of North

*Top:* A work boat and its master ply the water between Swan Point and Hart-Miller Island, spring 1972. (*News-American* Collection, Marylandia and Rare Books Department, University of Maryland Libraries)

*Bottom:* In 1984 the acreage on Hart-Miller had dropped to 176 (not counting Pleasure Island), but about ten years later, the Army Corps of Engineers selected the islands and the area south of them as a disposal area. By the early twenty-first century, the fill from Baltimore harbor had turned Hart-Miller into a park and refuge for water birds. (NOAA 12278, 2002. Courtesy of Maptech, Inc.)

Point State Park, which is a popular gathering place for boaters. In the summer, dozens of boats anchor off the three-thousand-foot-long beach. Many raft up for general fraternizing, and everyone enjoys swimming, eating, and crabbing.

## REFERENCES

1845 HC, USCGS, 166; 1846–47 TC, USCGS, 213 (Hart 197.11 acres, Millers 138.09 acres); 1854 TC, USCGS, 450; 1896 HC, USCGS, 2345; 1898 HC, USGS, 2399; 1897–98 TC, USCGS, 2308; TC, USCGS, 2326; 1944 Sparrows Point Q, USGS; 1984 NOAA 12278 (Hart 107.05 acres, Millers 30.58 acres, Pleasure Island 38.23 acres). Erosion loss Hart 0.64, Millers 0.78 acre per year.

"The Party Ships Out to Hart-Miller," *Sun,* Maryland Section, June 4, 1987.

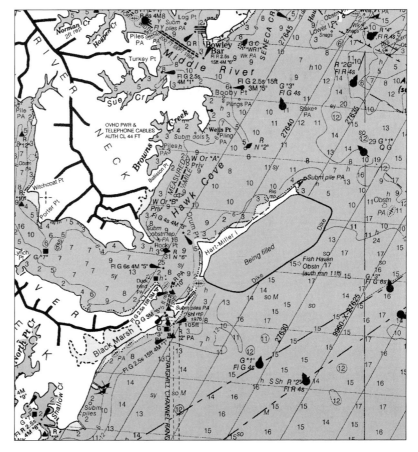

## FORT CARROLL
*Never a Shot Fired in Anger*

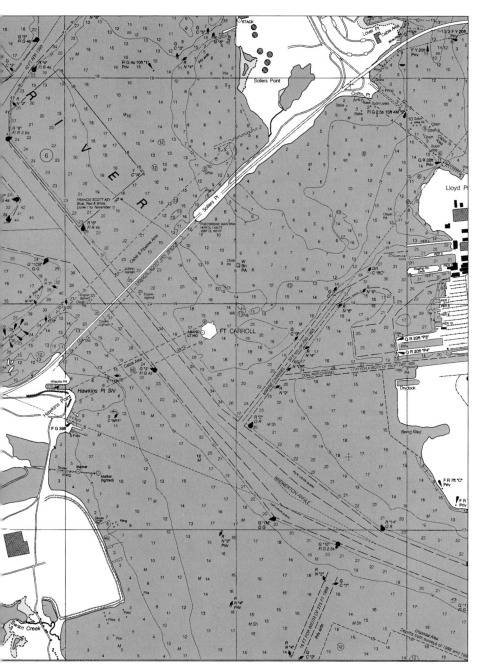

Fort Carroll—a defense post perfectly situated in the channel, now dilapidated but in outline unchanged since its construction—as it appears on a modern chart. (NOAA 12278, 2002. Courtesy of Maptech, Inc.)

Rock Point, and Fort Howard on North Point. The Army Corps of Engineers also designed and built Fort Carroll. Named for Charles Carroll, signer of the Declaration of Independence, the fort was built on a four-acre artificial island on Sollers Point Flats in the Patapsco River.

Preliminary surveys and other work on the fort began in 1847, and a year later an army engineer, Brevet Col. Robert E. Lee, arrived to take charge of the construction. The original plan called for a six-sided, four-story, forty-foot-high building designed with multi-tiered emplacements for 350 cannons. Each of the fort's six sides measured 246 feet long. Colonel Lee directed the work until 1851, when he was given command of the U.S. Military Academy at West Point. In the meantime Congress had put a stop to completion of the original plan. The fort did not go beyond the first tier.

The army decided that a light to aid in navigating the lower Patapsco was of greater value and in 1853 built a wooden lighthouse and a frame two-story keeper's house. The Fort Carroll light began operation in 1854, and the light keeper and his family became the first and only full-time residents on the fortified island. In 1898, during the Spanish-American War, the army tore down that first lighthouse and fog bell tower. One hundred feet north of the site, army construction crews built a new lighthouse, installing a new fog bell, lantern, and fifth-order Fresnel lens that stood forty-five feet above the water. On the walls of the fort were installed three gun batteries consisting of two twelve-inch guns, two five-inch guns, and two three-inch

Following the War of 1812, during which the British attacked Baltimore, the citizens demanded stronger fortifications against any future threat. As a result, the army refurbished Fort McHenry and built batteries at Fort Armistead on Hawkins Point, Fort Smallwood on

guns. During World War I, army gunners used these batteries for firing practice. Later, the army removed the large guns, and the light station was shut down in 1931. With the advent of World War II, Fort Carroll was again put to use, this time as a pistol range. The fort also served as short-term housing for arriving seamen while their ships were being fumigated. After the war, the army abandoned the island entirely.

Not until 1958 did anyone take an interest in the empty fort. That year Baltimore lawyer Benjamin Eisenberg bought the fort and island, rebuilt some of the bastions, installed wooden guns, and added a restaurant. He ran visitors out to the island by hydrofoil, but their numbers were never great, and Eisenberg had to give up his venture. Since

then entrepreneurs have proposed the island as a site for theaters, restaurants, marinas, a penitentiary, museums, and a gambling casino, but none have been built. The island remains a forlorn monument, now badly vandalized and deteriorating.

## REFERENCES

1866 HC, USCGS, 914; 1880 HC, USCGS, 1451; 1934 HC, USCGS, 5531; 1975–76 HC, USCGS, 9564; 1996 NOAA 12281.

Hornberger/Turbeyville. Lewis, Emanuel Raymond, *Seacoast Fortifications of the United States* (Missoula, Mont.: Pictorial Histories Publishing Company, 1970, 1979). Travers, Paul J., *The Patapsco: Baltimore's River of History* (Centreville, Md.: Tidewater, 1990).

# GIBSON ISLAND
## *An Exclusive Gated Community*

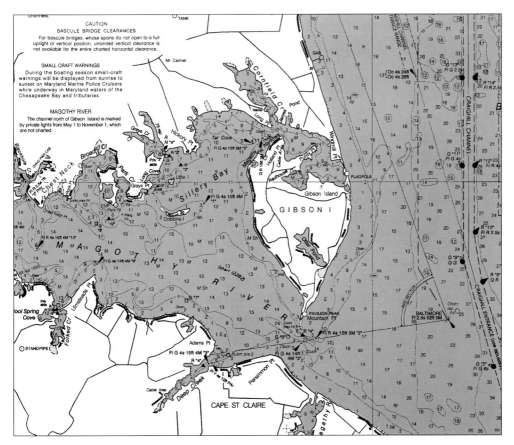

Time has treated Gibson Island comparatively well. In 1997 it retained about 900 of its original acres. (NOAA 12278, 2002. Courtesy of Maptech, Inc.)

At the mouth of the Magothy River, Gibson Island's nine hundred acres encompass a well-sheltered harbor, Otter and Cooley Ponds, and a very private community. Accessible by a single road and causeway that are closed to the public, the island community is guarded by a stone gatehouse that is manned twenty-four hours a day. The island's 180 or so property owners make up the Gibson Island Corporation, which manages the island. At the heart of this wealthy community is the exclusive Gibson Island Club. Members not only may partake of clubhouse activities but also may enjoy a golf course—eighteen holes if you play one nine in reverse—a pool, marina, and skeet-shooting range. However, members of the venerable Gibson Island Yacht Squadron and students at the Gibson Island Country School are not required to be island residents.

Gibson Island has had a long history of occupation. Artifacts and a large shell midden on the bay side of the island tell of Indian inhabitants who gathered oysters and other foods from the Chesapeake. These Indians were probably of Algonquian stock, pushed south by fierce Susquehannocks.

First patented in 1667, the island had four owners and contained approximately 1,200 acres. By 1771 William Worthington owned what was then known simply as "the great island," and in 1792 he advertised it for sale: "The land is remarkably valuable for the Fertility of its soil, and Conveniency of its situation to Baltimore, Annapolis, and other Markets . . . It abounds with Plenty of Timber and Wood. And has the advantage of productive Fisheries, and Variety of wild Fowl. It is adapted to every Species of Country-Produce. There are several different Tenements on it, with some useful Improvements, and it produces the greatest Plenty of Grass for Stock." In

1793 John Gibson, an Annapolis attorney, bought the island, which was part of the more than 2,000 acres of land he owned between the Magothy and Bodkin Creek.

Subsequent owners divided the island into farms, but the hilly, sometimes rugged landscape did not lend itself well to agriculture. Although during the Civil War its well-hidden harbor sheltered pro-Southern blockade runners, Gibson Island failed to attract later development. It was popular with deer and duck hunters, however. Finally, in 1921 Judge W. Stewart Symington Jr. headed a group of wealthy Baltimoreans who bought the island for $165,000. They brought in the nation's foremost landscape firm, the Olmsted brothers, who devised a plan for an elaborate summer colony with cottages organized around an eighteen-hole golf course. It was "Baltimore's newest playground," declared a 1923 article in the *Baltimore American*. Mostly woods and open space, the island was still a favorite with hunters, but Symington envisioned much more. He organized the Gibson

*Opposite:* In May 1938, the Baltimore *News-American* devoted a photo essay to Gibson Island, offering a glimpse of the enjoyable life it offered. The photographer captured golfers walking a plank pathway, sailboats anchored in the background. (*News-American* Collection, Marylandia and Rare Books Department, University of Maryland Libraries)

A patent dated 1667 conveyed 1,200 acres of what became Gibson Island, and a chart of 1844 portrayed it as almost entirely devoted to farming. (Courtesy of National Archives (NACP) RG 23, T175)

Members taking in the sun—while properly dressed—in front of the clubhouse. (*News-American* Collection, Marylandia and Rare Books Department, University of Maryland Libraries)

Island Club and had plans for the Yacht Squadron before his death in 1926. Today, a tall Celtic cross near the island's entrance honors his memory.

Following the Great Depression, real estate development made more economic sense than a golf-centered community, and the eighteen-hole course was reduced to nine holes. Gibson Island was too remote for year-round living, and development in the 1930s consisted of summer cottages. By the 1940s, however, families were taking up permanent residence on the island. The com-

munity experienced another spurt of development after World War II, and the increase in the number of young children on the island led to the creation of the Gibson Island Country School. Located outside the island gate, the school was accessible and open to children of nonislanders, who now make up most of the enrollees.

Since the 1920s the Gibson Island Yacht Squadron has flourished and earned wide acclaim. In 1929 J. Graham and C. Lowndes Johnson won the Star-Class Internationals, making the Gibson

Island Yacht Squadron the host for the championship races the following year. During World War II, a number of Gibson Island yachts served with the volunteer "Corsair Fleet" on Atlantic Coast antisubmarine patrol. Gibson Island resident Corrin Strong's cutter *Narada* was sunk off Cape Henry while on this duty.

Gibson Island's yacht squadron continued to host regattas and long-distance races for decades and offered a number of stellar sailing events. One involved the popular Star boats. An Olympic racing class of identical boats a little under twenty-three feet long, the Star boats attracted generations of good sailors, among them squadron member James Allsopp, who won the Star-Class World Cup in 1976. In 1990 the squadron hosted the J-22 World Championship Regatta, in which local sailor and author John Sherwood won first place. The squadron has continued to be active in racing and has grown to more than two hundred yachts. Some fifty Gibson Island children crew a junior fleet of Optimists, Lasers, and 420s.

A third generation of luxury homes appeared on the island during the area-wide real estate boom of the 1980s, and properties began to change hands, a thing that seldom happened previously. Nevertheless, residents attempted to uphold their unwritten policy of exclusivity. Although property ownership brings with it shares in the Gibson Island Corporation, club membership is not automatic, which is especially hard on island children barred from the pool, sailing, and other facilities. In 1991 the issue erupted in a legal battle when several property owners who were denied Gibson Island Club membership took the corporation to court. The final settlement granted all landowners use of the club's facilities but not club membership. While admitting no wrongdoing,

the corporation did pay the plaintiffs' $40,000 lawyer's fee.

Disruption of the status quo came from a different quarter in the mid-1980s. Hunting on the island had ended before World War II, and some time later, it became a bird sanctuary. Along with the birds, deer also flourished, eventually overrunning the lush Gibson Island neighborhoods. In 1984, the islanders requested that the state's forestry officials organize a deer hunt. Immediately picketers appeared at the gate to protest. They succeeded in postponing but could not stop the hunt, which is now an annual event, not only on Gibson Island but at Sandy Point State Park and similar sanctuaries.

The Gibson Island Historical Society maintains a small but nice museum that is open by appointment by writing to P.O. Box 667, Gibson Island, MD 21056.

REFERENCES

1667 Early patent (1,200 acres); 1933 HC, USCGS, 5403 (1,118.75 acres); 1845 HC, USCGS, 164 (1,116.64 acres); 1954 Gibson Island Q, USGS, (1,037.51 acres); 1997 NOAA 12252 (900.10 acres). Erosion loss 3.4 acres per year.

*Annual Report,* MdDNR, Wildlife and Heritage Division, 1997–98. *A Brief History of the Gibson Island Yacht Squadron* (Gibson Island Museum, n.d.). Brown, Johanne, "Island's Deer Hunt Picketed," *Annapolis Capital,* November 30, 1984. Casey, Dan, "Some Gibson Island Club Members Fighting Settlement," *Annapolis Capital,* June 24, 1991. Reynolds, Louisa B., *Gibson Island* (Baltimore: Hearsay, Reese Press, Inc., 1978). Rodgers, Patricia, personal communication, August 1998. Spencer, Duncan, "Gibson Island," *Annapolitan Magazine* (September 1989). Ware, Donna M., *Anne Arundel's Legacy: The Historic Properties of Anne Arundel County* (Annapolis: Environmental & Special Projects Division, Anne Arundel County Office of Planning & Zoning, 1990).

## DOBBINS ISLAND
*Deserted Island of Mystery and Memories*

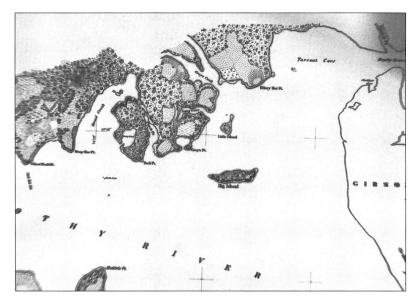

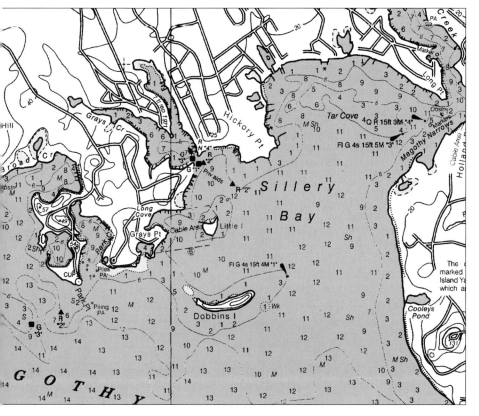

*Top:* Dobbins Island, 1844. Little Island is to the north. (Courtesy of National Archives (NACP) RG 23, T175)

*Bottom:* In a century and a half, Dobbins (privately owned and closed to the public) has lost more than half of its area. (NOAA 12278, 2002. Courtesy of Maptech, Inc.)

North of Gibson Island at the mouth of Sillery Bay, Dobbins Island is a sheltered Magothy River anchorage protected from storms and nasty weather. For years it has been a popular meeting place where yachtsmen rendezvous for parties on their boats.

One very early name for Dobbins Island is Dutch Ship Island, an intriguing name whose origins may go back to the first English settlers in Anne Arundel County. A group of Puritans from Virginia settled on the peninsula between the Severn and Magothy Rivers, founding Providence, one of Anne Arundel County's lost towns. They traded regularly with the Dutch, evidence of which was uncovered in archaeological explorations begun in the early 1990s by Anne Arundel County archaeologist Al Luckenbach. Numerous Dutch artifacts attest to the existence of the widely scattered settlement, which, like the Dutch ship, disappeared during the seventeenth century. Long before these and English settlers came to the area, Indians used Dobbins Island as a hunting camp, leaving behind arrowheads and other artifacts.

William Gambrel was the island's first known owner in 1769. At that time the island was recorded as comprising twelve acres. This is questionable because the first official survey, done in 1845, lists the island as being 18.34 acres. Throughout much of its history, Dobbins Island has been known as Big Island and has been paired with a tiny islet to its north, not surprisingly identified as Little Island. At one time the latter was also known as Raspberry Island.

In 1803 Dobbins Island, on which there was a working tobacco plantation, was deeded to Annapolis attorney John Gibson, who owned much of neighboring Gibson Island. In 1811 Gibson sold Dobbins Island to Richard Caton, and thereafter it passed through several hands. In the late 1800s George W. Dobbin bought it, and his name stayed with it when the Penniman family bought Dobbins and Little Islands in 1930.

According to T. Milton Oler, a resident of nearby Olmstead Point, Little Island underwent a temporary name change. For a time locals called it Palmer's Island after a family who owned a rambling old house on the island. The Palmers entertained often, and many of their guests arrived by crossing the sandbar on horseback. Eventually, a summer cottage replaced the Palmer house. Today Little and Dobbins Islands are still private property, and the owners discourage trespassing.

REFERENCES

1845 HC, USCGS, 164 (Big Island 18.34 acres; Little Island 4.63 acres); 1898 H2365 (Big Island 12.09 acres; Little Island 2.34 acres); 1933 HC, USCGS, 5403 (Big Island 8.77 acres; Little Island, 1.28 acres); 1998 NOAA 12282 (Dobbins Island 7.42 acres; Little Island 2.23 acres). Erosion loss 0.07 acre per year.
Luckenbach, Al, Ph.D., *Providence, 1649: The History and Archaeology of Anne Arundel County Maryland's First European Settlement* (Annapolis: Maryland State Archives & MdHT, 1995). Nelker, Gladys, *Town Neck Hundred of Anne Arundel County: The Land, 1649–1930* (Westminster, Md.: Family Line Publications, 1990). Penniman, Margaret, personal conversation, 1999. Taylor, Marianne, *My River Speaks* (Arnold, Md.: Bay Media, Inc., 1998).

## ST. HELENA ISLAND
*A Tiny Island with a Regal Mansion*

Embraced by the curve of Little Round Bay, nineteen-acre St. Helena Island is the setting for an impressive Federal Revival mansion, the centerpiece of an estate that occupies about six acres of landscaped grounds. The house was built between 1929 and 1931. Patterned after Homewood on the Johns Hopkins University campus in Baltimore, it was part of the early-twentieth-century classical revival movement in architecture.

A dream realized by Baltimore attorney Paul Burnett, the brick house is rich in detail, from its two-story rotunda to large elegant rooms with mahogany floors and marble fireplaces. Burnett commissioned bay schooners to bring in the necessary building materials. He chose the name St. Helena as a tribute to his former law partner, a great-great nephew of Napoleon Bonaparte, who spent the last years of his life in exile on the island of St. Helena in the South At-

St. Helena Island, 1844.
(Courtesy of National Archives (NACP) RG 23, T176)

lantic. Perhaps Burnett saw his island in a similar light.

Other buildings on the island served a variety of purposes over the years. Handicapped children and their counselors lived in several of them while attending summer camp. Burnett sold St.

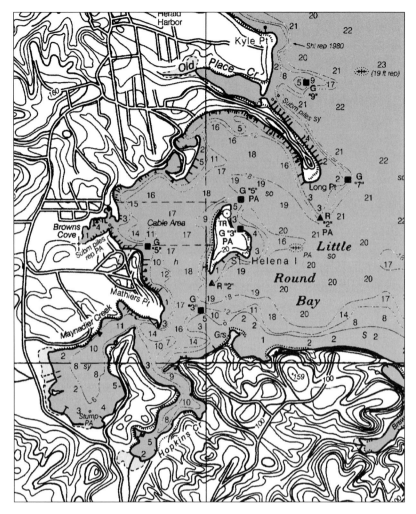

Sheltered St. Helena has lost only about 2 of its 21.5 acres in the last 150 years. (NOAA 12282, 2002. Courtesy of Maptech, Inc.)

Helena in the 1940s to Eugene Raney, who leased one of the buildings to Four Corners Restaurant owner John Emory. The building became a casino with slot machines that were seized in a 1951 police raid. The long pier that brought gamblers to the island still stands.

St. Helena Island was subdivided and sold again in 1956, passing through a number of owners. The six-acre estate with its house and gardens went through several stages of improvement, and realtor Chris Coile undertook a major restoration that brought the mansion back to its original splendor.

Centuries before St. Helena became a fashionable retreat, Woodland-period and later Indians occupied the site. They left behind evidence of their habitation,

including rough clay pottery shards, stone chips, and grooved axes. Legend has it that later tribes chose this and other small islands as places to hide from raiding Susquehannocks.

The island is also a refuge for wildlife, attracting a variety of ducks, geese, great blue herons, kingfishers, and other birds. Ospreys return regularly to nest, and there is a resident population of snakes and turtles. Once the island was graced by a deer park, but that has been overgrown for many years.

St. Helena Island and Little Round Bay are a part of two-mile-wide Round Bay, which is an unusual feature of the otherwise narrow and straight Severn River. The arcuate, or bowlike, shape of Little Round Bay is very similar to that of a meander in a river, which may explain its origins.

This idyllic setting created controversy in the 1990s when it attracted a new owner with big plans. Keith Osborne of Fantasy Island Management Corporation had a new life in mind for the estate, but his plans remained uncertain at the turn of the twenty-first century.

*References*

1844 TC, USCGS, 177 (21.53 acres); 1949 Anne Arundel County topographical map (nine buildings); 1981 NOAA 12282 (18.93 acres). Erosion loss 0.08 acres per year.

Davidson, A. Todd, and Colby Rucker, "Gems of the Severn," Severn River Commission, City of Annapolis (April 1988). "Fantasy Island Denied Private Club Status," *Annapolis Capital,* October 15, 1998. "For Sale: A $2.75 Million Island Retreat in the Severn River," *Annapolis Capital,* June 19, 1987. "Isles of Arundel," *Annapolis Capital,* May 14, 1987. "Maryland Inventory of Historic Properties," Site AA940, MdHT (n.d.). "No Man is an Island," *Annapolis Capital,* January 16, 1988. "Tiny Island New Home for Weddings?" *Annapolis Capital,* April 20, 1998.

Eastern Neck Island, near the mouth of the Chester River, is joined by bridge to the mainland. The entire 2,222-acre island is a national wildlife refuge.

Before the English settlers arrived, native peoples known as the Ozinies, or Wicomiss, lived in a village across the river and used the island as a base for shellfishing. Capt. John Smith noted their presence when he explored the bay in 1608. The Ozinies were hunter-gatherers who moved from place to place in search of food and other things they needed. They were not the first to explore the island, however. The recent discovery of a skeleton suggests a much earlier Indian presence. According to scientists in the Anthropology Department of Catholic University, the bones are the remains of a person who walked the island sometime around 1300 BC.

The island's recorded history begins about 1650 when Joseph Wickes came from Virginia to Maryland to become one of the earliest settlers on the Eastern Shore. By 1654 he owned 400 acres in Kent County, by 1656 he was married, and by 1659 he had acquired the southern half of Eastern Neck Island where he built a house known as Wickcliffe. He represented Kent County in the General Assembly and served many years as a justice of the Kent County Court, which occasionally met at Wickcliffe.

In 1852 the Wickes's island farm consisted of 375 acres and a new frame house. All that remains of the original Wickcliffe are a few bricks and a memorial tablet erected in 1975 to commemorate the birthplace of Revolutionary War hero Lambert Wickes. A mariner by trade, he was appointed a captain in the Continental Navy early in 1776. He took command of the newly commissioned

*Top:* Eastern Neck Island as depicted on a chart of 1846. (Courtesy of National Archives (NACP) RG23, T200)

*Bottom:* A National Wildlife Refuge since 1962 (along with South Marsh and Cedar Islands farther south), the island has lost 291 acres in the last 138 years. (NOAA 12272, 2002. Courtesy of Maptech, Inc.)

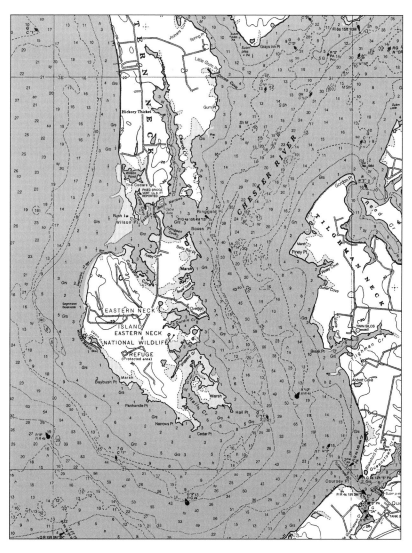

brig *Reprisal,* built originally as the merchant vessel *Molly.* Setting out on July 3, 1776, for his first mission, Wickes took the *Reprisal* south to the French Caribbean island of Martinique. At the entrance to the harbor, the American ship came up against the British sloop-of-war *Shark,* which blocked her way. With guns blazing the two ships fought the Revolution's first naval engagement in foreign waters. Half an hour after it began, the battle ended abruptly with neither vessel claiming victory. Angered by the British standing in the way of a ship trying to enter their port, the French fired on the *Shark* from their fort. The British ship sailed out of range, and the *Reprisal* continued into the anchorage. When he went ashore to call upon the French governor, Wickes claimed another first. He was the first American naval officer received officially by a representative of a foreign nation.

The following fall, the *Reprisal* sailed on a second mission, carrying Benjamin Franklin to persuade the French to enter the war on the American side. His charge delivered, Wickes stayed to raid British shipping, making him the first American naval officer to command an American warship in European waters. His orders to "cruise against our enemies" took him to the English Channel "to let Old England see how they like to have an active enemy at their own door." In January 1777 the *Reprisal* captured five vessels and headed with her prizes to the nearest French port. The English were incensed and immediately complained that the French were harboring "pirates and brigands." This was just the sort of ammunition that Benjamin Franklin needed. As relations between France and Britain worsened, the French edged closer to supporting the American cause.

By May 1777 Wickes was back at sea, this time with the former British customs cutter *Dolphin* and the brigantine *Lexington.* This small American squadron captured eighteen merchant vessels, sending alarm throughout British trading circles. Returning to France in June, the Americans mistook the seventy-four-gun ship-of-the-line HMS *Burford* for a merchant ship. Realizing that none of the American ships were a match for the British warship, Wickes ordered the other two to scatter. The largest of the three American vessels, the *Reprisal* immediately became the *Burford*'s prey in a chase that lasted for several hours. As the enemy ship closed in on the *Reprisal* off the coast of France, Wickes jettisoned his guns and anything else that would lighten his ship and increase her speed. Storm clouds darkened the skies, and the *Burford*'s commander ended the chase, choosing not to risk trouble so close to France's shores. Wickes and his ship made it safely to the port of St. Malo, where he later received orders to return to America. On September 14, 1777, the *Reprisal* set out across the Atlantic. Months passed, and news finally reached home that the ship had run into a storm off Newfoundland on or about October 1. The brig sank, taking Wickes and all but one of the crew with her. Had the ship's cook not survived to be picked up by a passing vessel, the disappearance of the *Reprisal* would still be a mystery.

The home in which Lambert Wickes was born was later enlarged and survived into the twentieth century to be torn down in 1935. On the site, J. Edward Johnston built a caretaker's cottage for the Cedar Point Gunning Club's lodge. Other parts of the island had also been developed. At the southernmost tip of Eastern Neck Island was Cedar Point Farm, originally a part of the Wickcliffe tract. There an early brick house was replaced around 1880 by a large frame house belonging to Benjamin and Frances Wickes Sappington. By 1928 the property had gone out of the Wickes family and into the hands of the Cedar Point Gunning Club, which demolished the Sappington house to build its hunting lodge.

The island's first settler, Joseph Wickes, had shared ownership of the island with Thomas Hynson, father of his third wife. Hynson bought the northern half of the island, which was identified as Ingleside when he passed it on to his son in 1680. A part of this property, known only as the Island Farm, belonged to David Jones in the late 1800s. Jones may have built the large brick house that survived into the 1960s. In the twentieth century the farm served mainly as a private hunting preserve.

Several other houses were built on the island over the years. According to Michael C. Bourne in his architectural study *Historic Houses of Kent County,* Spencer Hall was originally a two-story brick and clapboard house with a gambrel roof. It was part of Thomas Hynson's seventeenth-century holdings and remained in that family until the early 1800s. By the 1830s, the property and a much-enlarged house had passed from Richard Spencer II to his daughters Martha and Maria. Spencer Hall went with Maria when she married Alexander Harris, whose descendants owned it until 1940.

Simon J. Martinet's 1860 *Map of Kent County* shows tiny Cockneys Island southwest of Cedar Point and twenty-two farms or residences on Eastern Neck Island, as well as two schools. All of these have since completely disappeared, as have the hunting clubs and an oyster-shucking plant. On the southeastern tip of the island is Hail Point. Here, officials of the colonial government stopped all vessels entering the Chester River for quarantine and customs inspections.

In 1966 the U.S. Fish and Wildlife Service bought the entire island and created the Eastern Neck National Wildlife Refuge. Most of the surviving buildings were subsequently demolished. The island now is a major Upper Bay stopover for migrating wildfowl, including diving and paddle ducks, swans, geese, bald eagles, ospreys, herons, egrets, and many hawks and songbirds. The migratory waterfowl begin arriving in October, are most plentiful in November, and leave in early April. Among the animals that live on the island are white-tail deer, the endangered Delmarva fox-squirrel, rabbits, raccoons, opossums, muskrats, and skunks.

People are restricted to the six miles of roads and trails the Wildlife Service maintains and to an observation tower overlooking Tubby and Calfpasture Coves. These are a walk of about a mile from the landing at Overton on Durdin Creek. Formerly known as Bogle's Wharf, the landing served various packet vessels that regularly carried passengers and light freight from one part of the bay to another from colonial times until 1924. It is now a public launching facility open the year around. The refuge office has pamphlets on the various environments, the varieties of wildfowl, and information on permits for hunting, fishing, and using the boat launch.

REFERENCES

1846 MdGS (2,493.18 acres); 1846 TC, USCGS, 200; 1896 TC, USCGS, 2240; TC, USCGS, 2242; TC, USCGS, 2246; 1901 Chestertown Q, USGS; 1943 TC, USCGS, 8273; 1954 Langford Creek Q, USGS (2,291.2 acres); 1974 Langford Creek Q, USGS; 1984 NOAA 12272 (2,222.72 acres); MdIHP, K272 Ingleside, K273 Spencer Hall, K274 Wickcliff. Erosion loss 0.816 acre per year.
*Birds of Eastern Neck National Wildlife Refuge* (Washington, D.C.: U.S. Department of

the Interior, 1998). Bourne, Michael Owen, *Historic Houses of Kent County* (Chestertown, Md.: Historical Society of Kent County, 1998). *Eastern Neck Wildlife Refuge* (Washington, D.C.: Eastern Neck Wildlife Refuge, U.S. Department of the Interior, n.d.). Footner, Hulbert, *Rivers of the Eastern Shore* (Centreville, Md.: Tidewater, 1944; reprint 1989). *A Guide to Public Piers and Boat Ramps on Maryland Waters* (Annapolis: MdDNR, Tidewater Administration, 1999). McSherry, James, *History of Maryland* (Baltimore: Baltimore

Book Company, 1904). Meanly, Brooke, *Wildfowl of the Chesapeake Bay Country* (Centreville, Md.: Tidewater, 1982). *Trail of Life* (Washington, D.C.: Eastern Neck Wildlife Refuge, U.S. Department of the Interior, n.d.). Usilton, Fred G., *History of Kent County* (Chestertown, Md.: *Kent News*, 1980). Walsh, Richard, and W. L. Fox, eds., *Maryland, a History* (Baltimore: MdHS, Schneiderith and Sons, 1983). Wilstach, Paul, *Tidewater Maryland* (Cambridge, Md.: Tidewater, 1969).

## CACAWAY ISLAND
*Nearby Waters a Popular Anchorage, but Island Off-limits*

*Right:* View of Cacaway Island, 1846. (Courtesy of National Archives (NACP) RG 23, T201)

*Opposite page:* Cacaway Island, now private property and closed to visitors, has lost 2.5 acres from its 1846 area of 8.9 acres. (NOAA 12272, 2002. Courtesy of Maptech, Inc.)

Cacaway is a small uninhabited island in Langford Creek. Depending on the map used as a reference, the waterway also is referred to as Langford Bay because the creek widens into a small bay around Cacaway Island and then branches east and northwest. In 1672 the east fork of Langford Creek was called Steelpone Creek and the northwest fork, Jacobus Creek. Long and narrow, the island was once connected to the mainland. Topographic charts before 1897 show both a Cacawa Point and a Cacawa Island. The shape of Cacaway Island, as it has come

to be called, may suggest the origin of its name, which is obscure but could come from the Algonquian for porcupine quills.

The area was well populated by Indians before the English colonists arrived. Down the Chester River, the Ozinies had their main village. In 1670 cartographer Augustine Herrman indicated three Indian villages on the west side of what he called "Lanncefords Creek." Archaeologists have found several oyster-shell middens that the Indians left on the shores of Langford Creek.

In 1650 Lord Baltimore granted John Langford 1,500 acres, which he called Langford's Neck. Cacaway was undoubtedly a part of this grant. Later the area was called Broad Neck; it was subdivided in 1773 and leased to tenant farmers. Later transfers show the land changing hands at least nineteen times in the following years. Possibly the island was considered a part of Cacaway Farm, which was built at the southern tip of Broad Neck about 1810. In one of the few recorded mentions of Cacaway Island, it is said that sometime around 1860 James Haddaway lost the island in a poker game. Ultimately the Broad Neck tract was sold by David Wilmerding to the Cacaway Cooperative Housing Association in 1988. The 7.6-acre island was part of a parcel amounting to 183.2 acres that sold for $1.9 million.

Offering a sheltered anchorage and deep water close by, Cacaway Island is popular with yachtsmen. The entire island is private property, however, and sorties ashore are discouraged. Pets allowed to run free on the island are particularly unwelcome because of the threat they pose to the many songbirds that have found a refuge there.

REFERENCES

1670 Augustine Herrman Map, 1776 Anthony Smith Map, 1780 John Hinton Map, 1795 Dennis Griffith Map, Papenfuse/Coale; 1864 TC, USCGS, 201 (8.91 acres), 1877 Lake et al., *Atlas;* 1897–1905 TC, USCGS, 2291; 1933 Plat of Cacaway Farm & Island; 1937 TC, USCGS, 5699; 1942 TC, USCGS, 8273; 1984 NOAA 12272 (7.65 acres); 1998 NOAA 12272 (6.40 acres). Erosion loss 0.02 acre per year.

Bourne, Michael, *Historic Houses of Kent County* (Chestertown, Md: Historical Society of Kent County, 1998). *Guide to Cruising Chesapeake Bay* (Annapolis: *CBMag,* 1989). Macielag, Michael, interview by the author, Chestertown, Md., March 1989. Wilke, Steve, and Gail Thompson, *Prehistoric Archeological Resources in the Maryland Coastal Zone,* MdDNR, 1977.

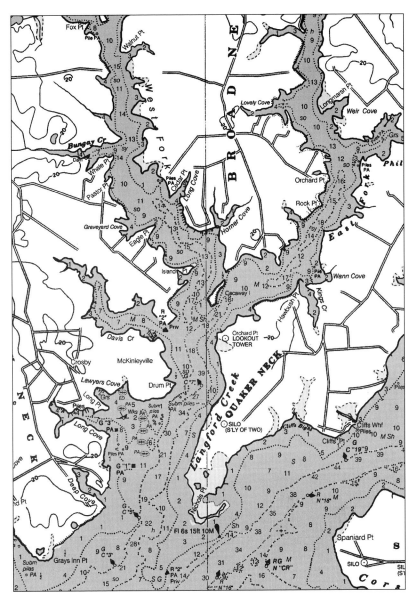

*Maryland's First Permanent European Settlement*

On August 17, 1631, Virginian William Claiborne, a partner in the London firm of Cloberry and Company, claimed a large Eastern Shore island in the middle bay for a settlement and trading post. At the time the English arrived, the island was occupied by Matapeake Indians who sold it to Claiborne for twelve pounds of trade goods. Naming it "Isle of Kent" after his birthplace, he chose a site east and north of Kent Point on the southern tip of the island and there erected a stockade protected by four cannons. About one hundred people made up this first permanent European settlement in what soon became the new colony of Maryland.

According to a brief history by Mildred C. Schock of the Kent Island Historical Society, during the first two

Kent Island depicted in the 1778 *Carte de la Baie de Chesapeake* by M. De Sartine, an early and not notably accurate chart. (Maryland Law Library).

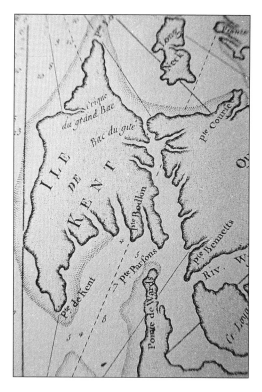

the southern tip of the island tells of trouble with the Matapeakes, who were lured to a conference and then slaughtered by colonists on what is known as Bloody Point. The colonists and Matapeakes apparently made their peace, and the Indians remained on the island until about 1770 when the last of them left their homes around the mouth of Broad Creek. What became of the various local Indian tribes is obscure—the Matapeakes, Mononposons, and Ozinies simply vanished from all known records.

Of greater concern to Claiborne and the Kent Island settlement were relations with the Maryland colonists who arrived in 1634. Under the charter given to Cecil Calvert, second Lord Baltimore, Kent Island was a part of Maryland. The fight between Claiborne and the Marylanders lasted more than a decade, during which Claiborne lost the *Long Tayle* in what some historians claim was the first naval battle in the new world. Certainly it was the first on the Chesapeake. After a long and bitter struggle, Claiborne lost all claim to the island.

The settlement on the island continued to flourish. In 1632 the Rev. Richard Jones built a church at Fort Kent, making its congregation the earliest organization in Maryland. This congregation was the core group that established Christ Church at Broad Creek in 1652. They later built a new church, where they worshiped until 1880 when they moved to yet another building in Stevensville.

By 1686 a settlement had grown up on Broad Creek, which became a stop on the postal route established by Maryland's General Assembly in 1695. Near it stood Workman's Inn, a tavern that served travelers using the Annapolis—Broad Creek Ferry. Broad Creek was also the location of the Eastern Shore's first courthouse and jail until the time of

years, Claiborne reported that the settlers built houses as well as a fort. Only recently have archaeologists discovered the long-submerged site of Fort Kent, where they found evidence of several wells, buildings, and many artifacts. Around the fort, the first settlers cleared land, planted corn, and tended to their hogs. Claiborne reported that he kept "men abroade in severall boates a tradeing which was our principall worke." He employed a shipwright who undoubtedly was kept busy building boats to handle the demands of his trade in fur and corn. Claiborne's pinnace, the *Long Tayle,* was the first boat built in Maryland.

Skilled in his dealings with the Indians, Claiborne controlled the fur trade with the Susquehannocks and other native peoples in the northern bay. Among those who lived on or around Kent Island, the Wicomiss and Nanticokes rose up against the settlers who had taken their lands. One of several early legends concerning the origin of the name given

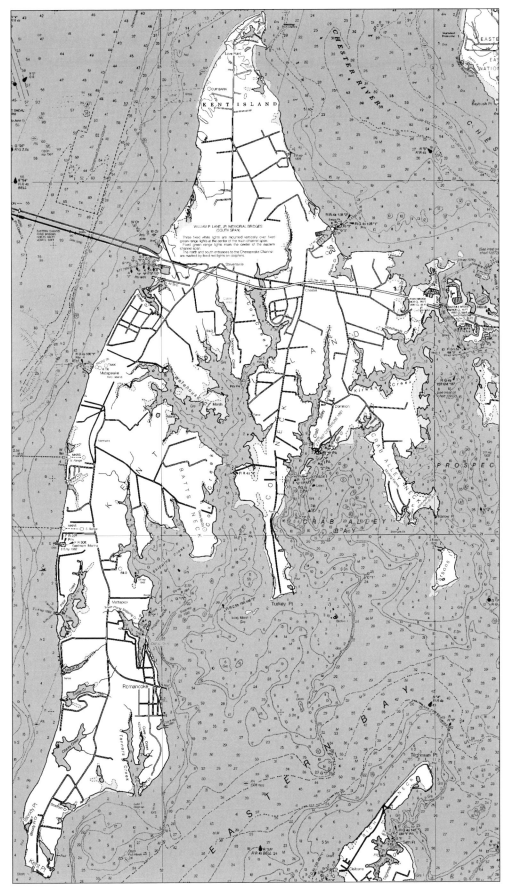

Kent, the largest island in the Chesapeake Bay, consisted of some 19,628 acres at the beginning of the twenty-first century, having lost an average of 12 acres per year since 1844. (NOAA 12263, 2002. Courtesy of Maptech, Inc.)

A deserted house near Romancoke, Kent Island, as the celebrated Maryland photographer, A. Aubrey Bodine, captured it in the mid-twentieth century.
(Bodine Collection, The Maryland Historical Society, Baltimore, Maryland)

the Revolution when the main north-south route shifted to the head of the bay. The Kent Island Historical Society has identified more than fifty historical homes and sites that help to tell the story of the island's development.

For many years the islanders were self-sufficient, supported mainly by farming. Tobacco was the chief crop until the 1800s, when farmers turned to wheat, corn, and, later, produce, dairy cattle, and poultry. Fishing and oystering fostered the growth of the water-oriented communities of Stevensville,

Dominion, and Chester. At one time, ten oyster-shucking houses operated on the island where fishing, oystering, and clamming are still important.

Until the first bridge spanned the bay in 1952, Kent Island's only ties with the Western Shore were by boat. Steamboats out of Baltimore made regular stops, and in 1902 the Pennsylvania Railroad ran a line to the steamboat landing at Love Point on the northern end of Kent Island. In later years the auto ferry *Philadelphia* ran between Baltimore and Love Point. She was a large, double-ended ferry whose telltale trail of black smoke earned her the nickname *Smokey Joe.* The hour-long ride across the bay was almost always a memorable one and, for people on their way to Ocean City, often ended with a pleasant stopover at the Love Point Hotel. It contributed to many happy memories until a devastating fire destroyed the building in the late 1940s.

Auto ferry service carrying commuters and vacationers between Annapolis and Claiborne on the Eastern Shore began in 1919 with the sidewheeler *Governor Emerson C. Harrington.* Eventually, the *Harrington* was replaced with the more modern, double-ended *Governor Albert C. Ritchie, John M. Dennis,* and *Governor Harry W. Nice,* and in 1930 the Eastern Shore end of the trip was moved to Matapeake on Kent Island.

All of this changed radically with the building of the Bay Bridge, especially the second span in 1972. At the peak of travel in August, more than 50,000 vehicles cross daily. Until the 1990s vacationers and other travelers crossed the busy Kent Narrows on the eastern shore of Kent Island by drawbridge, which on summer weekends created backups on land and water. All were speeded on their way with the building of the present span arching high above the waterway.

Kent Island has become home to a growing number of new residents who continue to contribute to the population explosion that began in the closing decades of the twentieth century. Acres of farmland, historic houses, and quaint villages have been eclipsed by housing developments, modern shopping centers, airfields, and marinas. At least fifteen new residential communities and attendant commercial growth have forever altered the face and economy of the island.

Nevertheless, traditional livelihoods continue to support a number of island families whose breadwinners still work the water. In 2000, Kent Island watermen brought in 476,350 pounds of fish, 16,645 bushels of oysters, and 301,683 pounds of soft- and hard-shell crabs.

Like every other island in the bay, this, the Chesapeake's largest island, is subject to erosion. Until the influx of new housing developments with their extensive bulkheading, the rate of erosion was two to four feet a year on the eastern side of the island and two to eight feet a year on the western side, which amounted to a loss of about twelve acres a year. Along shorelines lacking bulkheads, the loss to erosion remains the same and is a perpetual concern as development continues.

REFERENCES

1844 TC, USCGS 181; 1847 TC, USCGS 222; 1898 TC, USCGS 2316; 1898–1905 TC, USCGS 2325; 1973 PR Kent Island Q, USGS; 1974 PR Claiborne Q, USGS; 1977 Love Point Q, USGS; 1983 NOAA 12270; 1984 NOAA 1226619 (844.8 acres). Erosion loss approximately twelve acres per year.

Cronin, William, *Circumnavigating Kent Island* (unpublished manuscript, 1986). Denny, Emily Roe, *Indians of Kent Island* (Kent Island, Md.: Christ Church Parish and Kent Island Historical Tour Committee, 1959). Holly, David C., *Steamboat on the Chesapeake* (Centreville, Md.: Tidewater, 1987). Jensen, Ann, "Remembering the Annapolis Ferry," *Inside Annapolis Magazine* (December/January 1996/97). Schoch, Mildred C., *Of History and Houses, A Kent Island Heritage* (Queenstown, Md.: Queen Anne Press, 1982). Shomette, Donald G., *Ghost Fleet of Mallows Bay* (Centreville, Md: Tidewater, 1996). Wilke, Steve, and Gail Thompson, *Prehistoric Archeological Resources in the Maryland Coastal Zone,* MdDNR (1977).

## BODKIN ISLAND
### *From Dagger to Dot*

*Right:* "Pointe Bodkin," as depicted in M. De Sartine, *Carte de la Baie de Chesapeake* (1778). Note also the presence of "Pointe Parsons," like Bodkin at the time still attached to Kent Island (Maryland Law Library)

*Far right:* An 1899 chart of the island, which, since 1847, had eroded from about 50 acres to a little more than 32. (Courtesy of National Archives (NACP) RG 23, T2294)

*Bottom:* Bodkin Island, a speck in Eastern Bay, 2002. The state-owned island seemed likely to provide a disposal area when work began on the dredging of Kent Island Narrows Channel. (NOAA 12270, 2002. Courtesy of Maptech, Inc.)

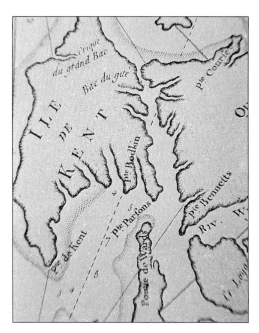

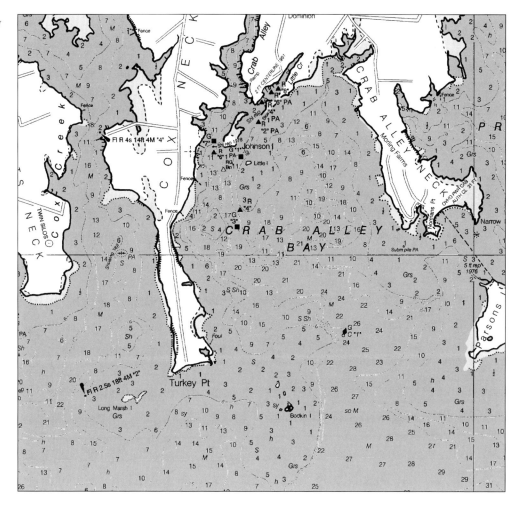

On modern charts, tiny Bodkin Island is little more than a dot off the lower end of Kent Island. Once, however, it was a peninsula of some significance, part of a land grant known as Sudlers Fortune. Charles Calvert, fifth Lord Baltimore, granted the tract, totaling more than 1,286 acres, to Joseph Sudler in 1742. Sudler, a Kent Island planter, died in 1756 and left 650 acres to his son Emory. The younger Sudler was also a planter as well as a merchant and delegate to the Maryland General Assembly.

Until the time of the American Revolution, Bodkin was a slim finger of land pointing across Eastern Bay toward the mouth of the Miles River. Early maps, produced in 1742, 1751, 1768, and 1776, show it still attached to the mainland. Long and narrow, the isthmus resembled a bodkin, which is a large blunt needle or a small dagger.

State records indicate that Emory Sudler fell on hard times. In 1775 he sold his Kent Island property to a Robert Anderson who paid 1,696 pounds sterling for it. The land eventually came into the hands of a Dekas Thompson who sold it to Gideon Emery in 1805. English and other foreign currencies were still legal tender in the United States, and when Emery made his purchase, he paid 184 pounds, two shillings for the property. Two years later, John Denny bought the spit of land with $2,150.

In 1847 four buildings—probably a house and barns, sheds, or other outbuildings—occupied the peninsula, which consisted of 50.03 acres, depending on the method of measurement. Apparently, tides and weather already were working on the isthmus. Local lore includes tales of residents driving their buggies through deepening water to

reach their homes. Not until 1864 did a map maker mention Bodkin by name and identify it as an island. Erosion continued to eat away at its shores, and in 1899 the island was down to thirty-two acres with little more than an orchard evident on a topographic chart of that year. Joseph Usilton, an early caretaker of the island, recalled fields of wheat surrounded by a forest of more than one thousand trees in which a great flock of herons made their homes. Just offshore, the waters were teeming with fish. Usilton remembers seeing as many as nine groups of seine fishermen working nearby waters with their nets.

In the 1930s a herd of free-roaming goats joined the herons as the island's chief occupants. Every autumn, ducks, geese, and other waterfowl visited Bodkin Island and eventually attracted a group of hunters. Calling themselves the Seven Island Duck Club, the hunters bought Bodkin in 1939. To save the island from erosion, they built a bulkhead, the first effort at conservation. Later the club also built a lodge, which undoubtedly attracted the next owners, the Bodkin Island Hunting Club.

Black ducks were the most plentiful of the waterfowl that drew hunters to the island. The U.S. Fish & Wildlife Service recorded seventy-six black duck nests on Bodkin Island as late as 1953. By then, the island had dwindled to five acres. It continued to erode and in 1984, a black-duck census showed only three nests on the island. Wild flowers still grew in profusion, however, and lured bright clouds of Monarch butterflies that stopped over on the island during their late summer migrations.

When, in 1982, the island went on the auction block, Washington attorney

Richard Earle bought it for $30,000. He made the island his summer retreat, converting the lodge to a vacation home, where family and friends enjoyed crabbing and fishing in nearby Crab Alley Bay. Earle immediately put $97,000 into bulkheading and enclosed 0.94 acres. He spent another $12,000 on sand fill and other improvements. Then came winter. Storms and high tides blew out the northeast corner of the bulkhead and much of the sand. The final destruction of Earle's retreat came in 1985 when vandals started a fire that swept the entire island. A single loblolly pine survived out of the many thousands that once covered the former peninsula. It stood a few years more, until a storm brought it down.

Although he could not prevent his island from washing away, Richard Earle did make an effort. Had he not, the entire island would have disappeared, like Sharps Island, becoming no more than a sandbar at low tide. Today the trees and the large heron rookery are gone, along with the ducks, butterflies, and seine-haulers who left with the declining fishery.

For a brief time, it looked like Bodkin Island had earned a reprieve when Maryland's Department of Natural Resources was looking for a place to dispose of the spoil from the dredging of the Kent Narrows Channel. The state bought the island in 1995 for $140,000, but the de-partment soon discovered that the project would be too expensive and abandoned it along with Bodkin Island. By 1997 an aerial photograph showed only a short stretch of surviving bulkhead and a few bushes on the deserted island.

## REFERENCES

1670 Augustine Herrman Map, 1689 John Thornton and William Fisher Map, 1692 Jacobus Robyn Map, 1751 Joshua Frye and Peter Jefferson Map, 1776 Anthony Smith Map, Papenfuse/Coale; 1768 Jeffries Topographic Map; 1847 TC, USCGS, 223 (50.03 acres); 1864 Col. Sir Henry James Map, Papenfuse/Coale; 1899 TC, USCGS, 2294 (32.30 acres); 1939 TC, USCGS, 5705; 1942 Kent Island Q, USGS (10.10 acres); 1973 *PR* Kent Island Q, USGS; 1983 NOAA 12270; 1984 MdDNR Shore Erosion Control Revetment Plan, 10-83 (49.09 acres). Average erosion loss 0.36 acres per year.

Emory, Frederick, *Queen Anne's County: Its Early History and Development* (Baltimore: MdHS, 1950). Goertemiller, Richard, "Cruise of the Month: Crab Alley Creek on Eastern Bay," *CBMag* 15 (May 1985). Gould, Clarence, *The Land System in Maryland 1720–1765* (Baltimore: Johns Hopkins Press, 1913). MdDNR Shore Erosion Control: SEC 10-83, February 1984. Stotts, Vernon, and Davis E. Davis, "The Black Duck in the Chesapeake Bay in Maryland: Breeding Behavior and Biology," Maryland Inland Game and Fish Commission, *Chesapeake Science* 1, nos. 3 and 4 (December 1960).

## PARSONS ISLAND
*Former Spice Island*

Boaters passing through Kent Narrows are familiar with Parsons Island, a ninety-seven-acre isle where Crab Alley and Prospect Bays meet at the northern end of Eastern Bay. Originally a peninsula known as "Parsons Neck" or "Parsons Point," the island's story is intimately connected to that of Kent Island.

Earliest records dating from William Claiborne's settlement of Kent Island in 1631 don't mention the peninsula by name. By 1649 Robert Vaughn, Maryland's commander of Kent Island, owned 500 acres, including the peninsula. At his death in 1668, Vaughn left the 200-acre point to his daughter Mary

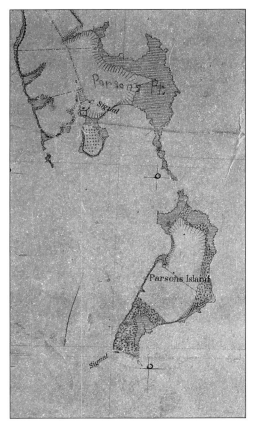

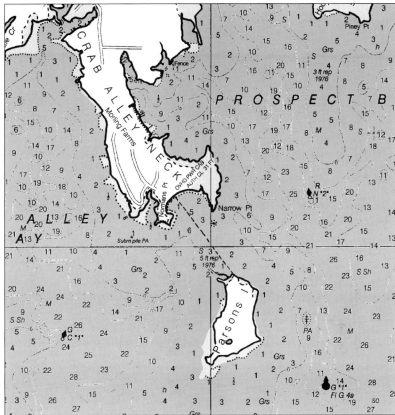

and her husband James Ringgold. The Ringgolds owned land on Parsons Point into the 1840s. In 1700 a John Parson owned land on Kent Island, but it is unclear if any of it was on the peninsula or if he's the Parson with whom the point was identified when it begins to be mentioned in records in 1725.

When the land became an island is difficult to say. Map makers disagree. Dennis Griffing's map of 1794 calls it Parsons Island but shows it attached to the mainland. Maps in 1795, 1833, and 1840 identify it as Parsons Point. Apparently, these early map makers did not verify details. Survey maps made in 1844 and 1847 finally show an actual island of about 190 acres separated from the peninsula by a channel one and one-half feet deep.

In the last quarter of the nineteenth century, Parsons Island had a number of different owners in rapid succession. An

1898 deed mentions a small stream separating the island from the mainland. An 1899 survey shows a four-foot channel, plus a house and several other buildings on the island. In 1904 a Baltimore wholesale liquor dealer, Thomas Ryan, raised racehorses and trained them on his own track on Parsons Island. He also had cattle and added a large barn. His widow sold the island to Charles Snyder for $3,510 in 1911, at which time it was possible to wade to the mainland. Snyder soon sold it to Findley French for five dollars plus considerations.

Joseph Usilton's father moved to the island in 1914 as overseer of the new farming operation. He, Joseph, and Joseph's brother, planted seven to eight thousand peach, apple, and plum trees. They also raised tomatoes, corn, chickens, and hogs and kept nine mules "for plowing and laying crops." During this time, the Duponts from Delaware

*Above, left:* An 1847 chart showed Parsons to be a wooded island of about 200 acres. (Courtesy of National Archives (NACP) RG 23, T223)

*Above, right:* Chart of Parsons Island, 2002. (NOAA 12270, 2002. Courtesy of Maptech, Inc.)

rented the island for hunting, and Joseph Usilton and his brother served as guides to the men who came down for a weekend of duck hunting. The men slept in a four-room cottage heated by a wood-burning stove. They took their meals, prepared by Joseph's mother and sisters, at the Usiltons' house.

In 1924 the Usiltons left the island when Findley French sold it to Jonathan Brown. The new owner reportedly crossed the channel to the mainland in a two-wheeled horse cart to buy his groceries, a practice he discontinued when "his groceries got wet." When Brown finally sold it in 1935, the channel between the island and mainland was half a mile wide.

The next owner, Chicagoan Willard Jacques, built a clubhouse, dredged and bulkheaded the harbor in an area known as Deep Creek, and constructed a boardwalk over the marsh. He sold the newly improved property to a Dr. Theodore Cooke. As the story goes, Cooke's caretaker had a going business making peach brandy. The good doctor regularly walked unknowingly within ten feet of the still until it was pointed out to him, and the brandy business came to an end.

Henry Breyer Jr., of Breyer's Ice Cream, bought the island in 1942 but sold it two years later to the McCormick spice company for a rumored $35,000. The official record shows it again was sold for five dollars plus considerations. At that point Joseph Usilton returned to Parsons Island. He and his wife moved into the old brick house that had been on the island for well over one hundred years. Later it was torn down, and they moved into a new house built by Mc-Cormick. Usilton and his wife were caretakers for twenty-three years.

According to the McCormick Company's history, because of product shortages and shipping problems during

World War II, they initially used the island as an experimental farm for growing spices and to test the company's pesticides on the crops. One year in the late 1940s, McCormick tried to raise turkeys on the island, but when Parsons Island Narrows froze just before Thanksgiving, and they could not get the turkeys to market, the enterprise came to an end. Cattle raising was similarly unsuccessful. Eventually, the company sold its pesticide lines, and most of the farm operation was shut down. McCormick left much of the island to indigenous wildlife and the growing of soybeans, corn, and grains.

When its agricultural enterprise ceased, the company focused on developing the island as a meeting and recreational facility for use by its executives, customers, business associates, and employees. Soon after its purchase of Parsons Island, McCormick built a clubhouse and cottages and over the years added modern conveniences. Phone service reached the island in 1950. Electricity was provided initially by three large generators and then by electric lines from the mainland. In 1970 the company replaced the cottages with an annex to the clubhouse.

Between April 1 and January 31, or until Parsons Island Narrows froze, the buildings were filled with people from all over the country. In the fall they hunted ducks, geese, and pheasant, and in the summer they took boats out onto nearby waters to fish. The company also maintained large motor boats, the *Island Queen* and the *Duchess,* to carry visitors to and from the island.

In 1989 ownership of Parsons Island passed from McCormick to Parsons Enterprises, a development company, but little change has taken place. Left largely to nature, the island has attracted a variety of wildfowl, especially black ducks. Mary-

*Opposite, top:* Aerial view of Parsons, 1986, when the McCormick spice company owned it and devoted much of it to crops. (Author photograph)

*Opposite, bottom:* Interior of the bright and comfortable Parsons Island Clubhouse (private), 1989. (Author photograph)

land's Department of Natural Resources recently recorded seventy-one widely scattered black duck nests, a few of which were in the unused duck blinds.

Erosion maps show that the western side of the island, particularly the southwestern corner that catches the brunt of summer winds and storms, is eroding at a rate of four to eight feet a year. The erosion rate for the remainder is two feet per year. Over the past 139 years, Parsons Island has lost a yearly average of four and one-half acres, and by 1997, erosion had whittled it down to 96.4 acres.

## REFERENCES

1670 Augustine Herrman Map; 1689 John Thornton & William Fisher Map; 1692 Jacobus Robyn Map; 1776 Anthony Smith Map; 1794 Dennis Griffith Map; 1833 David Burr Map; 1840 John Henry Alexander Map, Papenfuse/Coale; 1847 TC, USCGS, 223 (190.28 acres); 1850 Thomas Copperthwait & Co. Map, Papenfuse/Coale; 1899 TC, USCGS 2294 (157.19 acres); 1939 TC, USCGS 57051 (33.15 acres); 1973 *PR* Kent Island Q, USGS; 1983 NOAA 12270 (114.78 acres); 1997 NOAA 12270 (96.40 acres). Erosion loss 0.625 acres per year.

Isaac, Erich, *The First Century of the Settlement of Kent Island* (Ph.D. diss., Johns Hopkins University, 1957). Papenfuse/Coale. Shomette, Donald D., *Ghost Fleet of Mallows Bay* (Centreville, Md.: Tidewater, 1996). Steiner, B. C., "Kent County and Kent Island 1656–1662," *MdHM* 8 (1913). Stotts, Vernon D., and Davis E. Davis, "The Black Duck in the Chesapeake Bay of Maryland: Breeding Behavior and Biology," Maryland Inland Game & Fish Commission, *Chesapeake Science* 1, nos. 3 and 4 (December 1960). Thomas, Belvin B., interview by the author, March 1986. Usilton, Capt. Joseph, interview by the author, April 1986.

## POPLAR ISLAND
*Black Cat Farm, Presidents' Retreat, and Popular Resort*

The Poplar Island archipelago, 1862, oddly charted with north to the bottom. (Courtesy of National Archives (NACP) RG23, T215)

Today, much of Poplar Island is under water, emerging at low tide as seven separate land masses. Three actually have had names: Poplar, Coaches, and Jefferson. William Claiborne named the original island for one of his party of explorers, Richard Popeley, a man who thereafter passed into obscurity. By 1632 a Daniel Cugley was keeping a large herd of pigs on Popeley's Island and apparently supplied meat to Claiborne's settlement on Kent Island four miles to the north. Claiborne's records for September 1632 show that he paid two hundred pounds of tobacco for "a boare of Popeley's Island."

In 1631 Claiborne's cousin, Richard Thompson, with his wife and child had settled on the island, recorded as consisting of 1,430 acres. Thompson also transported seven indentured servants—one woman and six men—to work his plantation, which was a going concern until disaster struck in the summer of 1637. Thompson returned from a fur-trading voyage to find his wife and child and all of his servants murdered, his livestock slaughtered, and his house and farm buildings burned to the ground, reportedly by a band of marauding Nanticokes. Thompson later remarried and worked as an attorney. In the 1640s Thompson served in the Maryland General Assembly and as a justice and commissioner for Kent County, of which Poplar and Kent Islands were a part. His career in Maryland ended in 1644 when he took Claiborne's side during a revolt against Lord Baltimore's government and was declared an enemy of the province. He lost his land holdings and fled to Virginia.

The island's next owner was Thomas Hawkins, who sold half of it to his friend Seth Foster. Hawkins was a prosperous farmer when he died in 1656. His estate, which included a sizeable house, several outbuildings, farm equipment, servants, a boat, livestock, five paintings, a small library, a "Turke" carpet, and a looking glass, was worth 27,864 pounds of tobacco, a large amount for the period. Hawkins's widow later married Seth Foster. In 1669 they sold the island to Alexander D'Hinojosa, former governor of the Dutch colony of Delaware. Charles Calvert, third Lord Baltimore and governor of Maryland, had offered sanctuary to D'Hinojosa when the English government seized his colony near New Castle. He, his wife, and seven children lived on the island for many years.

Apparently during this time Popeley's became Poplar Island—perhaps because Richard Popeley was long forgotten, and a large number of tulip poplars flourished there. Cartographer Augus-

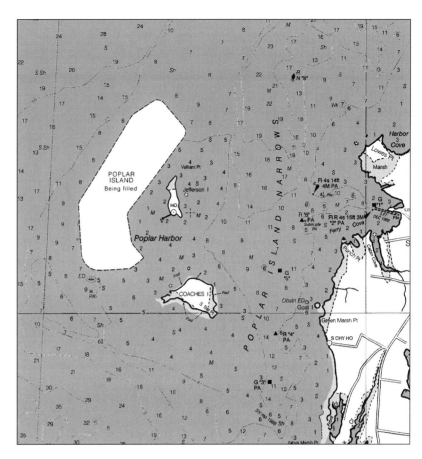

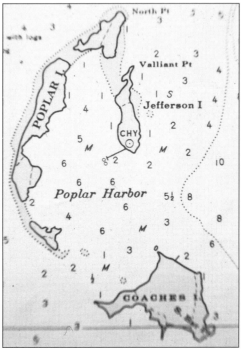

*Top:* Poplar Island, 2002. Between 1631 and 2000, when the state bought the island for use as a spoil-dumping site from the re-dredging of Baltimore harbor, Poplar's size dropped from 1,430 to 117 acres. (NOAA 12270, 2002. Courtesy of Maptech, Inc.)

*Bottom:* Well before 1973, rising water had broken Poplar Island into bits and pieces, the largest being Coaches Island to the southeast. (NOAA chart 550, 1973. Courtesy of Maptech, Inc.)

A clubhouse with a lovely knotty-pine-paneled interior stood on Jefferson Island, a member of the Poplar Island group, in the mid-1930s, when President Franklin Delano Roosevelt used the secluded spot as a getaway. Here FDR enjoys a moment on Jefferson Island with his guest, Postmaster General James A. Farley. The image appeared in a booklet celebrating the island's history, Peter K. Bailey, *Poplar Island: My Memories as a Boy* (Easton, Md.: Economy Printing, 1996), courtesy of Mary J. Fairbank.

tine Herrman made it official when he identified the island as Poplar Island on his 1670 map.

The island seems to have slumbered through the eighteenth century. A lone reference to it in 1781 records that the winter was especially severe. Ice six inches thick spread from the Upper Bay to the Potomac River, and islanders were able to cross the ice to and from nearby Tilghman Island in wagons and carriages.

Poplar Island residents' peaceful existence was shattered during the War of 1812 when crews from the invading British fleet occupied the island in the spring of 1813. While at anchor off the island, sailors caught "thousands of crabs" to supplement their diet. According to the island's owner, William Sears, who put in a claim for damages at war's end, they also helped themselves to "30 head of black cattle, 86 old sheep, 20–30 lambs, 300 breeding sows and pigs, and all the poultry that could be caught."

The next event of any note was a severe storm in December 1835 that caused

the loss off Poplar Island of the sloop *Hester Ann*. All aboard her were drowned, including John Paca, then owner of Wye Island.

Possibly the most remarkable piece of island history was the creation of the Great Poplar Island Black Cat Farm in 1847. Charles Carroll, grandson of Charles Carroll of Carrollton, one of the Maryland signers of the Declaration of Independence, owned Poplar Island. Responding to intelligence that China was a good market for the fur of black cats, Carroll decided to turn Poplar Island into a fur farm. In December of that year, Carroll's agent, R. O. Ridgeway, advertised that he wanted one thousand female black cats and would pay twenty-five cents each for them when "delivered at Poplar Island or my store." He hired a waterman to go out daily with a load of fish for the cats, who ran at large on the island. All proceeded well until the weather turned severely cold. The bay froze, and there was no getting a boatload of fish to the island. Without food, the hungry cats took off over the ice, and that was the end of the black cat farm.

By the late nineteenth century, erosion had radically changed the face of Poplar Island, making of it what appeared to be three islands. To the south was Coaches Neck, connected by a narrow strip of land visible at low tide. To the east, across the harbor—known locally as "the pot"—was Cobbler's Neck, later renamed Jefferson Island.

In the 1880s about eighty-five people lived on the three islands, which still supported several farms. In the winter several men harvested oysters from local waters. Island watermen made large catches that sold for thirty-five to fifty cents a bushel. The rest of the year they fished, crabbed, and grew wheat and tomatoes, which were then taken by freight schooners to Baltimore markets. Poplar Island had its own post office, a sawmill, a general store, and a school that doubled as a church on Sundays. Mainland doctors made regular island visits.

By the turn of the century, many people had moved away. A 1900 map shows only four houses. The few remaining children attended classes in one of the houses, and the school building was used strictly for Sunday services. At some point during the 1920s, the last permanent residents left Poplar Island, which became a hunting preserve. Ignoring Prohibition laws, moonshiners used the islands for their illicit purposes until 1929, when the Talbot County sheriff brought in three revenue men who arrested five moonshiners, destroyed a thousand-gallon still, and dumped 21,500 gallons of whiskey.

That same year, a group of prominent Democrats bought Poplar and Jefferson Islands, a purchase heralding the beginning of the island's most illustrious period. As quoted in Peter K. Bailey's history of Poplar Island, the politicians sought "a quiet, undisturbed and attractive spot where (we) might mix the travail of political conferences with the pleasantries of clubhouse fraternity and where the humdrum of party politics might be broken now and then by communion with the great outdoors." Espousing the principles of Thomas Jefferson, they established the exclusive Jefferson Islands Club in 1931. In recognition of the island group's new status, the Maryland legislature renamed it Jefferson Island. Many of the nation's most prominent Democrats and businessmen weekended at the club, often called "a shrine for Democrats." The presidential suite was occupied regularly by Presidents Roosevelt and Truman while Secret Service men prowled the island and the navy patrolled offshore. The era ended when the clubhouse burned to the ground in 1946. Two years later George and Marion Bailey bought Poplar and Jefferson Islands and built the Poplar Islands Lodge on the site of the old Democratic Club. The hunting and fishing club opened in time for the 1948–49 hunting season, and Peter Bailey, who was a boy at the time, remembers "guests came from all over the East Coast, from Maine to Florida," and from midwestern states, too. George Bailey died in the summer of 1951 at the age of forty-four, and in September Marion Bailey wrote "The End" to Poplar Islands Lodge and her family's pleasant life there.

The islands passed through several hands before being sold in 1967 to the Smithsonian Institution, which used them for a variety of projects. Researchers studied heron and osprey populations and their breeding patterns, and the differences between island and mainland populations of the Carolina wren. They conducted small mammal studies, delved into the ecology of small forest patches, and monitored chemical pollutants.

For many years ospreys nested on the island, and avian census takers counted as many as thirty nests—one of the largest concentrations in the world. The ospreys arrived about the second week in March—on St. Patrick's Day according to local fishermen—and departed in September for their wintering grounds in Brazil, Venezuela, and Colombia. The islands were also a roost for large flocks of crows, great blue herons, and egrets, but the continual erosion has caused a significant loss of habitat, severely diminishing these populations. Gone completely are the tulip poplars that gave the island group its name.

By the 1990s, according to Peter Bailey, Poplar Island itself had almost completely disappeared. When William Claiborne explored it in 1627, it consisted of 1,500 acres. All that remains now are four small bits of land totaling no more than five acres. These shoals have helped to keep Coaches and Jefferson Islands from suffering the same fate, although they were down to less than 118 acres in 1997.

At the turn of the twenty-first century, the state of Maryland owned Poplar Island and used it for dredge disposal from the shipping channels into Baltimore. The twenty-year-long $427-million project will eventually create a new island approximating the shape and historical acreage of the original. Once again, it will provide a habitat for ospreys, blue herons, egrets, and other wildlife.

REFERENCES

1631 survey (1,430 acres); 1846 TC, USCGS, 215 (Poplar 633.31 acres, Jefferson 22.91 acres, Coaches 161.64 acres, total 817.86 acres); 1877 Lake et al., *Atlas;* 1888 TC, USCGS, 2293 (Poplar 362.75 acres, Jefferson 22.91 acres, Coaches 125.37 acres, total 511.03 acres); 1904 Annapolis Q, USGS; 1942 Claiborne Q, USGS (three islands total 292 acres); 1972 MdDNR Survey TA-1RL-45 & 46 (Poplar 35.3 acres, Coaches 140.97 acres, six islands total 261.2 acres); 1985 MdDNR 147-0334 (Poplar 7.34 acres, Coaches approximately 135.00 acres, Jefferson 17.91 acres, six islands total 166.9 acres); 1997 NOAA Chart 12266 (seven Islands 117.96 acres at low tide). Erosion loss 0.3 acres per year.

Abb, Irving, personal communication, April 1985. Bailey, Peter K, *Poplar Island, My Memories as a Boy* (Easton, Md.: Peter K. Bailey, 1996). Blanchard, Fessenden S., *A Cruising Guide to the Chesapeake* (New York: Dodd Mead, 1950). Craven, Avery O., *Soil Exhaustion as a Factor in the Agricultural History of Virginia & Maryland, 1606–1860* (Urbana: University of Illinois Press, n.d.). *Guide to Cruising Chesapeake Bay* (Annapolis: *CBMag,* 1998). Hayes, Anne, and Harriet Hazleton, *Chesapeake Kaleidoscope* (Cambridge, Md.: Tidewater, 1975). Higman, Daniel, *A Short History of Poplar Island, Talbot County,* Chesapeake Center for Environmental Studies, Smithsonian Institution (n.d.). Kinsman, Dorothy, "The Bay's Disappearing Island," *Sun Mag,* August 20, 1970. "Land Notes, 1634–1655," *MdHM* 6 (1911). Papenfuse/Coale. Preston, Dickson, *Talbot County, a History* (Centreville, Md.: Tidewater, 1983). Richardson, Hester, *Sidelights on Maryland History* (Cambridge, Md.: Tidewater, 1967). Semmes, Raphael, *Captains and Mariners of Early Maryland* (New York: Arno Press, 1937; reprint, Baltimore: Johns Hopkins University Press, 1979)

A beautifully secluded Eastern Shore retreat of nearly 2,000 acres, Wye Island
and the nearby Conference Center of the
Aspen Institute were much in the news
in the late 1990s as the site of peace talks
between Israeli and Palestinian leaders.
The island, which is surrounded by the
Wye and Wye East Rivers and Wye Narrows, is now a natural resource area.
Spanning the narrows is a ten-foot-high
bridge that connects the island and
mainland. Without a draw, it prevents
boats needing greater clearance from
circumnavigating the island.

Arrowheads, pottery shards, and bits
of stone tools found on the island attest
to the presence of local Algonquian
tribes who preceded a line of illustrious
colonial owners—Lloyds, Bordleys, and
Pacas being the most notable. The first
names connected with the island, however, were those of the five English
patent holders in the early 1600s:
William Stevens, Thomas Clary, William
Clary, William Price, and Thomas Bradnox. The only place name that has survived from those early claims is Drum
Point, part of Thomas Bradnox's patent.
In 1668 Thomas Clary's plantation, The
Purchase, was the only property that was
not bought by Philemon Lloyd. The
only son of Edward Lloyd, he was the
founding member of a wealthy and po-
litically powerful family that played a
significant role in the first 150 years of
Maryland's history. Philemon represented Talbot County in the General
Assembly, establishing a political legacy
that later yielded governors and U.S.
senators.

On the eve of the American Revolution, two leading Maryland patriots
shared ownership of Wye Island: John

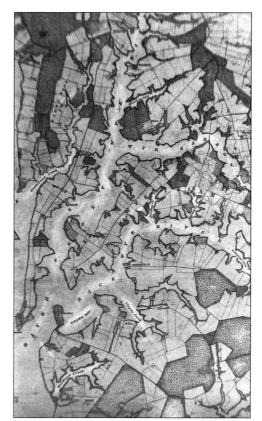

*Top:* Wye Island, highly cultivated, as it appeared in a
chart of 1847. At the dawn
of the twenty-first century,
the state of Maryland
owned nearly all of Wye Island, preserving its 1,900-
odd acres (it loses about five
acres a year) as a Natural
Resources Wildlife Management Area. (Courtesy of National Archives (NACP) RG 23,
T213)

*Bottom:* Wye Island, 2002.
(NOAA 12270, 2002. Courtesy of
Maptech, Inc.)

The magnificent Paca house, Wye Hall, once described as the most splendid house in America, 1989. (Author photograph)

Beale Bordley owned the western half from Dividing Creek to Bordley's Point; William Paca, the eastern half.

Bordley acquired the island when his first wife, Margaret Chew, inherited it from her father, who had married into the Lloyd family. A passionate agronomist, he seized the opportunity to establish an experimental farm. Bordley regularly tried out new crops and agricultural methods, the first of which was to replenish the depleted land. From the mid-1600s it had been planted with tobacco, severely draining the soil of its nutrients. On Wye Island Bordley put his theories on the benefits of "due tillage, proper rotation of crops . . . and manures" into practice. By rotating his principal crop, which was wheat, with others, he began to see results. He also imported cattle and sheep from England and practiced scientific breeding.

The goal of Bordley's Wye Island ex-periment was to develop a completely self-sufficient operation. Although he owned slaves, he condemned the institution, believing that it was not good for agriculture because it made a farmer "a slave to his slaves." Eventually, he either freed many of his slaves or bound them out as apprentices. Most likely, slaves and free blacks supplied the labor for Wye Island farm's complex operation. To provide salt for his livestock, Bordley used a rough desalinization process to extract it from the Wye River. He set up a system for threshing wheat using four six-horse teams on a circular wheat-strewn track. Each team circled the track at a walk for six laps, trotted six, and then rested as another team took its turn. Bordley also had a kiln to make bricks, made his own gunpowder; and grew flax and cotton, which was then woven into cloth colored by dyes from his own plants. He grew hemp to make his own ropes, raised bees for honey, and planted a large orchard of figs, plums, pears, and pomegranates. He had his own brewery and his Tokay grapes were widely praised for their flavor. During the Revolution, Bordley's island enterprise helped support the American cause. He provided gunpowder, wheat, and meat for the Continental Army.

A colonial legislator, one of four Maryland signers of the Declaration of Independence, and a postwar governor of Maryland, William Paca also obtained his Wye Island property through marriage. His first wife, Ann Mary Chew, was Margaret Bordley's sister. Before the Revolution, Paca maintained a large tobacco plantation. With the coming of the war, he helped supply the Continental Army with beef and wheat from his farm. After the war, he took a more personal interest in the island. There, in 1791, he built Wye Hall, viewed by one historian as "the most splendid house in America," and hired a professional landscape architect to design its gardens. William Paca died at Wye Hall in 1799. Tragedy later struck the Pacas, whose family was divided by the Civil War. William B. Paca of Wye Hall was a Northern sympathizer, whereas Edward Paca, who lived at Wye Plantation on the mainland, sympathized with the South. In a feud over land, William Paca was accused and subsequently acquitted of killing two members of Edward's family. The sad story did not end there, however. William died soon after the trial, two of his sons committed suicide, a daughter died of accidental poisoning, and in 1879 Wye Hall burned when a workman repairing the roof accidentally sparked a fire.

In the 1920s multimillionaire Glenn Stewart bought a significant portion of the island. Unlike John Beale Bordley, he knew little about farming and ulti-mately failed at raising Percheron horses, sheep, and Hereford cattle. Stewart enjoyed hunting, however, and built a gunning lodge, which he called the "Duck House." As the story goes, he once lived in Guatemala where he killed two intruders whom he caught robbing his home. Told that the Guatemalans were intent on "blood revenge," he left the country and subsequently built a fortress-like house on the Miles River near Wye Island. His Duck House was of similar impregnable construction with steel plates on the doors and windows, and a large basement with a secret entrance under the main floor fireplace. A hydraulic lift lowered the section of floor to allow access. Stewart, it seems, preferred the seclusion of his private yacht, which set off with him to Nassau one day, never to return. When his widow, Jacqueline, died in 1964, she left a tangled multimillion-dollar estate. In the basement of Duck House, her executors found bushel baskets and grain sacks full of jewelry and silver coins.

Following an unsuccessful attempt by the Rouse Company to buy the island and transform it into a modern residential community of estates and a central village, Maryland's Department of Natural Resources acquired the island and created the Wye Island Natural Resource Management Area to be administered by the Maryland Forest, Park, and Wildlife Service. The Duck House became a conference center. Although allowed by permit, goose and deer hunting on the island is strictly controlled. Shotguns, bows and arrows, and muzzle loaders have their own seasons. Goose pits are available but limited. Bordley would have been gratified to know that most of the island is still used for farming, with four large parcels rented for that purpose every year. Some of the land remains in private hands.

REFERENCES

1670 Augustine Herrman Map, 1795 Samuel
    Lewis Map of Maryland, Papenfuse/Coale;
    1847 TC, USCGS, 223, TC, USCGS, 224;
    1942 TC, USCGS, 8267, TC, USCGS,
    8268; 1942 Wye Mills Q, USGS; 1942 St.
    Michaels Q, USGS; 1942 Queenstown Q,
    USGS; 1983 NOAA 12270 (2,749.76
    acres); 1998 NOAA 12270 (1,986.56). Ero-
    sion loss 50.8 acres per year.

Arnett et al. Emory, Frederick, *History of Queen
    Anne's County* (Baltimore: MdHS, 1950).
    Gibbons, Boyd, *Wye Island* (Baltimore:
    Johns Hopkins University Press, 1977).
    Preston, D. J., *Talbot County, A History*
    (Centreville, Md.: Tidewater, 1963). Pa-
    penfuse et al. Weeks, Christopher, *Where
    Land and Water Intertwine: An Architec-
    tural History of Talbot County* (Baltimore:
    Johns Hopkins University Press and
    MdHT, 1984).

## BRUFFS ISLAND
*Private Island with Links to Early Colonial Port*

At the junction of the Wye and Wye East Rivers on Maryland's Eastern Shore, Bruffs Island extends a protective arm along the eastern rim of Shaw Bay. The resulting anchorage is large, open, and blessed most summer evenings by cool breezes. It can easily accommodate an entire yacht club fleet, making it a popular stopping-off place.

The record of English occupation of the island goes back to October 18, 1658, when Lord Baltimore granted colonist Henry Morgan 300 acres, known as Morgan St. Michaell. The grant included the forty-acre island, which was then separated from the mainland by a deep strait of water. Morgan later sold the island, along with another sixty acres or so to a William Crouch. Known for a time thereafter as Crouches Island, the entire one hundred acres, identified as Crouch's Choyce, passed at William's death to his son, Josias.

In 1683 when Maryland's General Assembly passed the "Act for the Advancement of Trade," it designated the town land at Wye River, variously called Wye Town and Doncaster, as an official port of entry. Such port towns were laid out in one hundred one-acre lots, with open spaces for markets, churches, or public buildings. The port of Doncaster was on the mainland adjacent to the island.

Around 1665 Thomas Bruff, a former London silversmith, bought the island along with a parcel on the mainland. At his death in 1702, he left his Doncaster plantation and half of Crouches Island to his son Richard and the remaining half of the island to his son, Thomas Jr., whose son, Richard Bruff, was an innkeeper in the town and also owned a large tobacco warehouse fronting on the stream between the island and the mainland. The foundation of the warehouse was unearthed in 1912 when digging began for construction of a seawall and the roadway that now connects the island to the mainland. In Bruff's day the stream was deep enough to handle freight schooners loading tobacco. Apparently in the 1690s, shipments in and out of Doncaster fell prey to pirates with enough regularity for the town's businessmen to petition Maryland's General Assembly for protection. By 1700, however, the town was able to deal with miscreants, boasting a whipping post and a pair of stocks erected by Richard Bruff for carrying out punishments mandated by the local court.

The town prospered, and fifty new lots were added in 1706, but that was also the year when Queen Anne's County was created from a part of Talbot County. The town fathers applied to have Doncaster named the new county seat, but that honor fell for a short time to Oxford and then to the village that became Easton. Doncaster's prosperity was headed for a decline in 1707 when Thomas Bruff Jr. sold what was by that time identified as Bruffs Island to Edmund Lloyd. Doncaster's status as a business center continued to diminish, although as late as 1780, Thomas Ray and William Sherwood were appointed as tobacco inspectors at Bruff's warehouse.

In 1888 John Howard Lloyd built a house that remained in the Lloyd family until 1909. Sidney Schuyler bought the island, adding a causeway to the mainland in 1914. The next major change came with the island's purchase in 1974 by Henry M. Witt and his wife. They built a new brick house, landscaped the grounds, and added 1,775 feet of stone revetments to protect the island from erosion. Today, the island remains private property.

REFERENCES

1794 Dennis Griffith Map, 1797 D. F. Sotzman Map, 1833 David H. Burr Map, 1833 Henry S. Tanner Map, 1840 John Henry Alexander Map, 1850 Thomas Copperthwait & Co. Map, Papenfuse/Coale; 1847 MdGS (45.92 acres); 1847 TC, USCGS, 223 (38.62 acres); 1877 Lake et al., *Atlas;* 1900 TC, USCGS, 2584; 1942 TC, USCGS, 8258; 1942 St. Michaels Q, USGS; 1997 NOAA 12270 (22 acres, two islands). Erosion loss 0.16 acre per year.

Mullikin, James C., *Ghost Towns of Talbot County* (Easton, Md.: Easton Publishing Company, 1951). Papenfuse/Coale. Tilghman, Oswald, *History of Talbot County, Maryland* (Baltimore: Williams & Wilkins Co., 1915; reprint, Baltimore: Regional Publishing Company, 1967). Weeks, Christopher, *Where Land and Water Intertwine: An Architectural History of Talbot County* (Baltimore: Johns Hopkins University Press and MdHT, 1984).

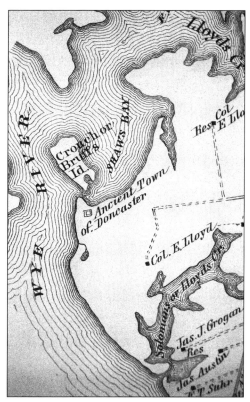

*Top:* Crouch or Bruffs Island as Lake, Griffing, and Stevenson's *Atlas of Talbot County* depicted it in 1877. A narrow channel separated the island from the mainland and the abandoned town of Doncaster. (Courtesy of Anne Arundel County Public Library)

*Bottom:* Bruffs lost half its area in 150 years, eroding from 46 acres in 1847 to 22 in 1997. (NOAA 12270, 2002. Courtesy of Maptech, Inc.)

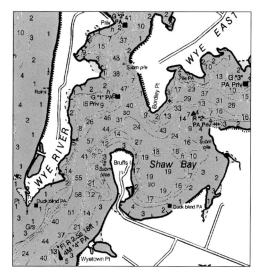

# THE MIDDLE BAY
## FROM THE CHOPTANK RIVER TO POCOMOKE SOUND

Human occupation of the islands of the Middle Bay goes back thousands of years. The early native peoples, followed by Choptank, Nanticoke, Manokin, Patuxent, and Annemessex Indians, used these islands mainly as sites for hunting and fishing camps. Stone tools, bits of pottery, and other artifacts attest to their presence on islands that once consisted of far more land than they do today.

The Indians reached the islands in dugout canoes that caught the attention of the Chesapeake's first English settlers, who borrowed the idea to fashion the hull of a sailing craft that came to be known as the log canoe. Many of these unique boats were built at island shipyards, where their design underwent further development. The result was the *bugeye,* a classic Chesapeake Bay sailing dredge boat, the last of which worked into the 1960s.

As elsewhere on the Chesapeake, Middle-Bay islanders were caught up in events of the Revolutionary War. The British raided Taylors, Hoopers, and other islands for supplies. Spring Island, just east of Holland Island, may have been the site of temporary British fortifications. Throughout the war and even after the fighting stopped, loyalists threatened island communities. Loyalists were especially active around Holland and Hoopers Islands. On occasion their depredations were so bad that Maryland's General Assembly sent troops to islanders' defense. In the case of patriot Matthew Tilghman, an armed Maryland naval vessel patrolled off Tilghman Island to protect his property. Islanders were not merely passive victims, and many fought back locally or joined the Continental Army, serving with the famed Maryland Line.

The War of 1812 again brought the British to prey on islands of the Middle Bay. They used Watts and other islands as bases from which to send out raiding parties and control shipping in the Tangier Straits.

Island life was not measured by war's consequences, which usually had no long-term effect. Islanders were used to trials and hardships, and human enterprise continued to flourish where the will and conditions were right. Boatbuilding history was made in the yards on Tilghman, Hoopers, Deal, and Solomons Islands, and the seafood industries sustained large numbers of island watermen and their families as they worked the water to supply world markets. That energy led to conflicts that occasionally reached the proportions of war in the last third of the nineteenth century. Dubbed the "Oyster Wars," the confrontations between oystermen of

*Previous pages:* Waiting at Solomons for the Baltimore steam boat, about 1920. (H. Graham Wood Collection, Mariners' Museum, Newport News, Va.)

Maryland and Virginia, and between oystermen and Maryland's "oyster police," who strove to curb the lawlessness and bloodshed, turned the waters of the Middle Bay into a battleground.

The period of the Oyster Wars was a sad and angry time with many tales of derring-do, but there is also a very different anger and sadness to stories of the Middle Bay islands. Many were heavily populated with homes, schools, churches, and stores. Harbors sheltered large fleets of working sail and power boats. For some, however, the good life couldn't last. Land washed steadily away, threatening homes and other buildings and forcing the residents to abandon their islands. Many took their homes, outbuildings, and even fences with them when they left for safer islands or the mainland. Holland, James, Little Deal, Janes, and Watts are now the exclusive domain of wildlife, ranging from native song birds, great blue herons, bald ea-gles, egrets, and seasonal migrations of waterfowl to small animals, goats, and even Sitka deer.

Tilghman, Taylors, Hoopers, Solomons, and Broomes are still well populated with thriving water-oriented communities and a number of traditional industries, such as crabbing, fishing, and boat building. Solomons Island is also the site of an important Chesapeake Bay resource, the Calvert Marine Museum, as well as the University of Maryland's Center for Environmental Sciences. Great Fox Island has become a vital educational center operated by the Chesapeake Bay Foundation. Several of these islands of the Middle Bay have, quite literally, become classrooms and laboratories where professional and amateur scientists—and great numbers of children and adults—may study and learn about the Chesapeake Bay and better understand the forces at work in this vast water-wrapped environment.

## Tilghman Island
### Rich in Bay Boat-building and Maritime History

Tilghman Island is a low sandy expanse of land about three and one-half miles long and from one-half to more than a mile wide. Travelers approaching via Maryland Route 33 cross Knapps Narrows on a new bridge completed in 1999. Rotting foundation pilings threatened the stability of the old bridge, which had stood since 1934. It was moved to the Chesapeake Bay Maritime Museum in St. Michaels, where it serves as an entrance gate. Watermen and pleasure boaters use Knapps Narrows to shave miles off their transit between the Choptank River and the Chesapeake Bay.

Relics found around Pawpaw Cove suggest that the island may have sup-ported an Indian village or a hunting or fishing camp site. During his exploration in 1608, Capt. John Smith made note of the island and the narrow channel separating it from the mainland. Augustine Herrman included both island and channel on his 1670 map of Maryland. Seth Foster, an Englishman, was the first person connected to the island but, it seems, only on paper. Although he received a grant that included it in 1659, there is no record that Foster actually lived on the island.

In 1662 Robert Knapp owned a small piece of land on the island side of the narrows, which still bears his name. Several others owned the island, but the most significant was Matthew Tilghman

*Top:* Tilghman Island, 1877, according to Lake, Griffing, and Stevenson's *Atlas of Talbot County.* The cartographers identified many dwelling places on the island, particularly at its upper part, which bore the name Bay Hundred. (Courtesy of Anne Arundel County Public Library)

*Bottom:* Tilghman Island on a contemporary chart. The island has lost 672 acres over the last 150 years, averaging 4.4 acres per year. (NOAA 12266, 2002. Courtesy of Maptech, Inc.)

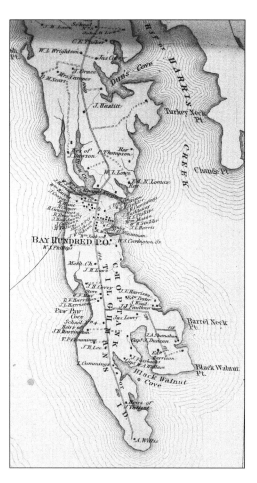

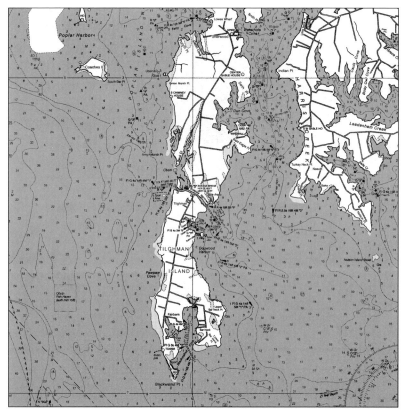

Ward, who acquired it in the mid-1700s. An attorney, legislator, justice, and planter, Ward married into the prominent Lloyd family. He and his wife, Margaret, had one daughter. Ward adopted his first cousin, Matthew Tilghman, and made him principal heir to his estate when he died in 1741. Tilghman's inheritance included the island, and under his ownership the first bridge was built to span Knapps Narrows. Not surprisingly, the island was named for Matthew Tilghman, who was a legislator, a delegate to the Continental Congress, and president of Maryland's senate during much of the Revolutionary War. Referred to as the "Father of the Revolution" in Maryland, he was largely responsible for the orderly way in which Maryland's government made the transition from a colony to a state. Because of his stand for Maryland's independence, Eastern Shore loyalists threatened to seize Tilghman's property, which brought the Maryland navy's barge, *Experiment,* to patrol the waters off Tilghman Island. Loyalist activities didn't cease when the war ended in 1781, and the *Experiment* continued to patrol the waters of Eastern Bay until 1782.

During the War of 1812, British troops used Tilghman's Island as a supply base. They carried off forty-three cattle, fifteen calves, fifty sheep, twelve barrels of corn, and three tons of hay belonging to resident Alexander Helmsley. Thus supplied, British forces went on to attack and burn Washington.

In the early 1800s two large steam sawmills operated on the south end but were not replaced when they burned. Later, as the island's population grew, several wind-driven grist mills handled the grinding of local grain. By 1862 boat building and oystering brought people to the new towns of Tilghman on the island side of the Narrows and Fairbank

*Top:* In the early years of the twentieth century, a long wharf extended from the south end of the island into the Chesapeake Bay, providing the residents a lifeline to the outside world. (H. Graham Wood photograph, Mariners' Museum, Newport News, Va.)

*Bottom:* An L-shaped frame house on Tilghman—a design apparently unique to the island. (Author photograph, 1966)

and Bar Neck at the southern end. Seven years later a new bridge crossed Knapps Narrows, but improvements didn't extend to island roads, which remained quagmires until 1898. That year construction began on the island's first oyster-shell roads, a successful solution used by communities all over the Eastern Shore.

Steamboat service came to the island with the first steamboat wharf built in 1892 and continued into the 1920s. In 1921 Tilghman Island residents could take an overnight steamer to Baltimore, leaving at 11:30 P.M. and arriving bright and early the next morning for a day in the city. They then boarded a steamer at

Baltimore's Light Street Wharf at 5 P.M. and were home by 10 that night. After its stop at Tilghman, the steamboat continued up the Choptank.

As elsewhere on the bay, water was almost always the better way to travel. Not long after the first English settlers arrived, they adapted the Indians' dugout canoe to their own uses. As time passed, they gave it a more boatlike shape, sharpening bow and stern, increasing its size by adding a second log, and adding sails. Tilghman Island was one of several places on the bay that became known for the building of "log canoes."

By the time of the Civil War, working watermen, like those on Tilghman Island, needed larger, heavier vessels to dredge oysters. To meet the need, builders refined the design of the log canoe, using five to nine or even eleven logs to make a larger boat and increasing the number of sails. The result was known as the bugeye, a handsome and excellent sailing vessel. When large trees grew scarce, builders retained the bugeye's basic hull lines but depended more on plank and frame construction. No one knows for sure how these boats came to be called bugeyes, but it was a common term by the 1870s. Some say it came from the word *buckeye,* since the hull of a bugeye looked something like a section from the outer shell of a horse chestnut. According to bay historian Robert H. Burgess, others claimed it came from the Scottish word *buckie* for oyster shell. The last bugeye working on the bay was the *Edna E. Lockwood,* built at Tilghman in 1889 by John B. Harrison for Daniel Haddaway. She ended her career as a work boat in the late 1960s and is now preserved at the Chesapeake Bay Maritime Museum in nearby St. Michaels.

Tilghman Island boasted several well-known builders of bugeyes, pungies, sloops, schooners, and other bay craft. Fishing and other work boats built on the island, as elsewhere on the Shore, can be distinguished by their graceful lines flaring gradually from a high bow to a square stem. With plenty of working room aft, they please both commercial and sport fishermen. This has not changed with the coming of the power boat.

Steam power in the 1800s and then the gasoline engine in the early 1900s gradually, but relentlessly, put an end to working sail. Many watermen simply abandoned their boats on river banks; others converted them to power. Beginning in the 1890s, the economical skipjacks replaced bugeyes and schooners in the oyster fleet. In the 2000–01 oyster season, thirteen skipjacks made up the last sailing fleet working in North America. The largest group of working skipjacks still sails out of Tilghman.

The majority of the island's 750 residents still make their living from the water, and today Tilghman is the primary seafood center in Talbot County. In a recent year, island watermen hauled in 1,959,649 pounds of fish, 6,737 bushels of oysters, and 659,254 pounds of soft- and hard-shell crabs. Sport fishing plays a more important role now than in former days, and the island is home to a large fleet of charter fishing boats. Below the narrows on the eastern side is Dogwood Harbor, location of the popular Harrison's Chesapeake House restaurant and fishing center. Southeast of Dogwood Harbor, at the end of an oyster-shell causeway is Avalon Island, the nineteenth-century site of the Tilghman Packing Company. Founded by brothers S. Taylor and J. C. Harrison, the oyster- and fish-packing house originally rose above the water on pilings, but over the years the great volume of discarded shells built up an island around the building and contributed to the causeway to the mainland. North of Avalon was a smaller island built by John B. Harrison. Also created from oyster shells, this island was not connected to the shore, and oyster shuckers traveled back and forth by skiff, an unpleasant and even dangerous business during winter gales. That may have caused it to be called Devil's Island, a place few mourned when it eventually washed away.

Along the island's developed waterfront, protective bulkheading has stopped significant erosion. In 1846 the island covered 1,905.85 acres, but over the last 150 years, storms and tides have accounted for an average of 4.4 acres per year, particularly on the bay side and southern end. In the same period of time, silting has decreased the depth of the narrows to five feet. With a mean tide range of 1.3 feet, a tidal current of up to two knots, and a system of channel markers that reverses at the bridge, boaters must navigate the narrow waterway with caution.

Hunting and fishing seasons are busy times on Tilghman, which is a starting point for hundreds of sportsmen who arrive by boat, launch their boats at Dogwood Harbor, or take one of the many charter boats. They find plenty of dockage, marine services, restaurants, lodging, and atmosphere in this historic watermen's village.

REFERENCES

1775 Tilghman survey (1,468 acres); 1846 HC, USCGS, 188 (1,905.85 acres); 1900 TC, USCGS, 2513); 1904 Sharps Island Q, USGS; 1942 Tilghman Island Q, USGS; 1943 HC, USCGS, 6958 (1,470.95 acres); 1971 DNR (1,264.75 acres); 1983 NOAA 12266 (1,261.75 acres); 1998 NOAA 12236 (1,237.58 acres). Erosion loss 4.40 acres per year.
"Bay Claiming Bits of Prehistory at Tilghman Island," *Sun*, May 10, 1998. Burgess, Robert H., *Chesapeake Sailing Craft: Part 1* (Cambridge, Md.: Tidewater, 1975). Emory, Frederick, *Queen Anne's County, Maryland: Its Early History and Development* (Baltimore: MdHS, 1950). "Explore Tilghman Island in the Water and on Bicycle," *Sun*, August 16, 1992. Preston, D. J., *Talbot County, A History* (Centreville, Md: Tidewater, 1983). Sinclair, R. R., *The Tilghman's Island Story, 1659–1954* (private printing, 1954). "Tilghman Island: Place Out of Time," *Sun*, September 14, 1997. "Time and the Tide of Development Arrive at Tilghman Island Despite Reservations," *Sun*, July 15, 1991. Willis, Dale, "Time, Tide Don't Wait," *Sun*, July 1, 1997.

## JAMES ISLAND
*Once Flourishing Community Now Domain of Sitka Deer*

James Island lies just north of Taylors Island in the mouth of the Little Choptank River. Within historical times, it was attached to Taylors Island, but the "great washout," which began in the nineteenth century, left more than one-half mile of open water between them. First settled in the early 1660s, St. James Island, as it was originally called, totaled 1,350 acres and consisted of at least three separate grants. Eventually, a family named Pattison bought the entire island and maintained ownership for more than two hundred years.

By 1877, maps showed two islands totaling 1,134 acres that encompassed more than a dozen homes, a school, and a store. The storekeeper was Capt. James T. Leonard, one of three owners of the Chesapeake Bay sloop *J. T. Leonard,* which was built on nearby Taylors Island in 1882 by Moses H. Geoghegan. She outlived her namesake and was still dredging long after relentless erosion

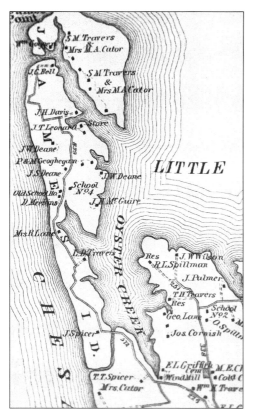

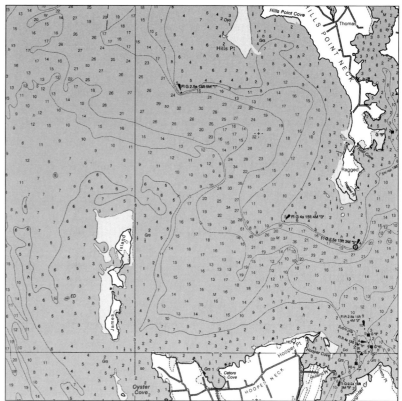

*Opposite, top left:* James Island, a detail from Lake, Griffing, and Stevenson's *Atlas of Dorchester County* (1877). (Courtesy of Anne Arundel County Public Library)

*Opposite, top right:* As surveyed in 1650, James Island consisted of 1,350 acres. By 1994 it had eroded to three small islands, totaling about 85 acres. In the early twenty-first century, no people but some Sitka deer lived on what remained of James Island. (NOAA 12266, 2002. Courtesy of Maptech, Inc.)

*Opposite, bottom:* Built on Taylors Island in 1882, the sloop-rigged skipjack *J. T. Leonard* proudly carried the name of a James Island store owner. The boat was demolished in 1974. (Author photograph, 1959)

*Above:* Aubrey Bodine visited James Island in 1958, when he photographed a man standing on—perhaps repairing, though it would appear not to require much work—a plank road made of downed timbers. The reason for the road and its maker supply fuel for speculation. (Bodine Collection, The Maryland Historical Society, Baltimore, Maryland)

had claimed Captain Leonard's store and the land on which it had stood. The *Leonard* was given to the Chesapeake Bay Maritime Museum at St. Michaels in 1968 and was dismantled in 1974.

A Methodist church was built on James in 1881, and in 1892 twenty families lived on the island. By 1910 the hardships of island life had driven all but seven residents to the mainland, and these soon followed. Cambridge resident Clement Henry found new occupants for James Island when in 1916 he imported a herd of Sitka deer. The tiny elk, standing only two-and-a-half feet tall at the shoulders and weighing from sixty to eighty pounds, were prized in their native Japan for their antlers, which were believed to produce a powerful aphrodisiac. As a result, the deer were hunted almost to extinction until the Japanese government stepped in and made them a national treasure.

The imported deer throve and multiplied on James Island where indigenous whitetail deer eventually joined them. The combined populations survived despite severe winters like that of 1957, when 161 deer perished from starvation. Maryland's Department of Natural Resources now maintains herds of Sitka deer on James and Assateague Islands and allows hunting to keep their numbers at levels that will prevent a depletion of their food supply. In recent years, the popularity of Sitka deer hunting has

increased, with more than 10,000 hunters taking advantage of each year's seven-day hunting season on the two islands. In Dorchester County, hunters took 632 antlered deer in 1995–96, 668 in 1997–98, and 547 in 1998–99. The 1997–98 harvests of stags and antlerless deer were nearly equal.

As with other animal and human populations on bay islands, the deer may lose their habitat as rising waters, erosion, and damaging waves continue to eat away at James Island at a rate of eight acres a year. Present charts show that James has become three small islands totaling 85.14 acres.

REFERENCES

1660s original survey (1,350 acres); 1847 TC, USCGS, 250; 1849 TC, USCGS, 272; 1877 Lake et al., *Atlas* (two islands, 774 acres and 360 acres, total 1,134 acres); 1900 TC, USCGS 2494 (two islands total 1,042.42 acres); 1998 NOAA 12261 (Upper 58.16 acres, Middle 0.22 acres, Lower 26.76 acres, total 85.14 acres) Erosion loss eight acres per year.
Badger, Curtis J., "Little Elk on the Chesapeake," *CBMag* (1990). *Game Program: Annual Report*, MdDNR, Wildlife and Heritage Division, 1997–98. Geohegan, Philmore W., personal letter, June 10, 1986. Gillmer, Thomas C., *Chesapeake Bay Sloops* (St. Michaels, Md.: Chesapeake Bay Maritime Museum, 1982). Hunter, J. Fred, "Erosion and Sedimentation in Chesapeake Bay Around the Mouth of the Choptank River," *General Geology*, U.S. Geological Survey (1914). Mowbray.

TAYLORS ISLAND
*Island of Pirates, Patriots, and Shipbuilders*

Northern Taylors Island forms the lower lip of the mouth of the Little Choptank River. From there it stretches southward about sixteen miles and is separated from the rest of Dorchester County by Slaughter Creek. Much of the lower is-

land is a watery maze that divides Taylors into several small islands shaped by meandering creeks and great expanses of marshland. Descriptive names offer clues to the nature of the various waterways. In addition to creeks and narrows, there

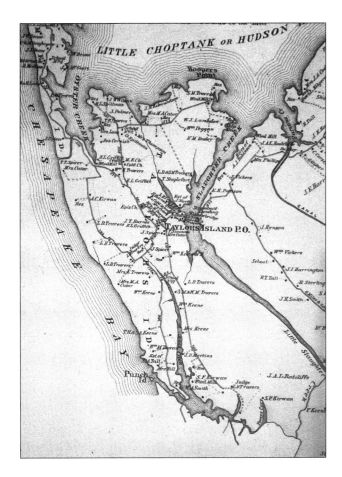

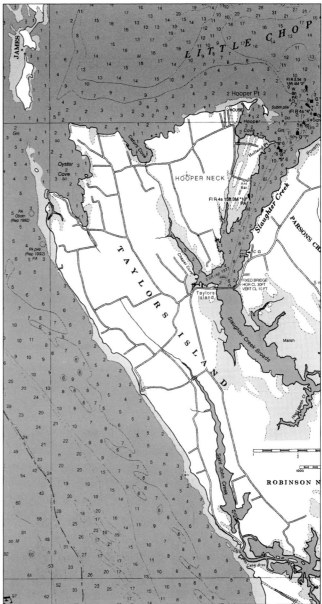

are places called broads, where a water-way expands; guts, meaning small, narrow creeks; and sloughs, which are marshes and swamps. Slaughter Creek, for instance, runs into Slaughter Creek Broads, then Slaughter Narrows, then Hog Marsh Guts, then Upper Keene Broad, and then Punch Island Creek, which flows into Dunnock Slough. The one landmark that is not a waterway is the two-hundred-year-old Neild Oak that stands near the middle of the island.

Most residents live on the upper island in the town of Taylor on Slaughter Creek. As the story goes, the creek got its name from a man who joined the hated seventeenth-century pirate Rodger Makeele, whose exploits are chronicled in Donald G. Shomette's *Pirates on the Chesapeake.* In the 1680s Makeele sailed out of Watts Island, preying on local shipping and the few unfortunate people

who then lived on the shores of the bay. Punch Island Creek acquired its name some 250 years later but from similar illicit activity. Punch Island, which lies between the creek and the bay, was said to have been a landing place for bootleggers during Prohibition.

Colonists who came across the bay from St. Mary's and Calvert Counties in 1659 were the first to settle the island. Primarily farmers, they raised tobacco and corn in the rich "Elkton silt loam." The name Taylor in association with the

*Top, left:* Taylor Island as it appeared in Lake, Griffing, and Stevenson's *Atlas of Dorchester County* (1877). In the late nineteenth century the island was quite well populated. (Courtesy of Anne Arundel County Public Library)

*Top, right:* By 2002 Taylors Island, consisting of some 8,077 acres in 1877, had eroded to 7,600 acres, a loss of about 4 acres a year or 5 percent over more than a century. (NOAA 12264, 2002. Courtesy of Maptech, Inc.)

island appears as early as 1662. Taylors owned land there for three generations. Philip was followed by Thomas and John. All three held the office of sheriff, and Thomas and John served in the legislature following the formation of Dorchester County in 1668. The oldest deed in county land records dates to 1669 when Thomas Taylor and his wife Frances sold a 1,200-acre tract called "Taylors Inheritance" to an Arthur Wright. A captain in the county militia, Thomas Taylor led a force that subdued a group of rebelling Nanticokes in 1678. The Maryland assembly reimbursed Taylor and fellow island militiamen Raymond Staplefort, Richard Tubman, and William Robson for expenses incurred during the campaign. His leadership abilities apparently earned Taylor a promotion to major two years later.

Another prominent islander was Thomas Pattison, who came to Dorchester in 1671 and was a county commissioner and clerk of the court. He may have built the first section of the main house at Mulberry Grove Plantation on the south side of Oyster Cove.

The house with its eighteen-inch-thick walls is the center section of the rambling building that stands on the spot today. This house was inhabited by Pattisons and Spicers until recently. The plantation's name came from the mulberry trees brought from Asia by Richard Pattison, who hoped to raise silkworms. Ultimately his venture was a failure, but it was responsible for a brief period of wild speculation in mulberry trees.

The first island church was built near Oyster Creek in 1709. It was moved to the grounds of the Grace Protestant Episcopal Church on Hooper Neck Road in 1959, and there it was restored. On its original site, the building may also have served as Dorchester County's first schoolhouse.

In 1728 the people of Taylors Islands petitioned the justices of Dorchester County for a ferry between the island and the mainland. They cited a population of 236 souls living on Taylors Island and their need to attend church as well as do business on the mainland. The justices granted one thousand pounds of tobacco to anyone who would establish

A Taylors Island landmark, Patrick's Discovery dates from 1870 or so and boasts interior woodwork of exceptional quality. (Author photograph, 1983)

the ferry. No records can be found of the ferry's first or earliest operators, but in 1786 Henry Travers held the position, and later, his widow.

During the American Revolution, officers and soldiers from Taylors Island served in the Continental Army. They were members of the Sixth Independent Maryland Company under Capt. Thomas Woolford and a "Flying Camp" under Capt. Thomas Bourke. Among four hundred Marylanders, they stood their ground against a British force of ten thousand seasoned troops at the Battle of Long Island in August 1776. There, 256 Maryland men died and were

buried in a common grave in Brooklyn. Throughout the war, George Washington depended upon the men of the "Maryland Line," and the record of their valor earned Maryland its nickname "The Old Line State."

In 1775 Capt. Benjamin Keene of Taylors Island organized a local militia battalion. Five years later Capt. John Brohawn reorganized the battalion, which was assigned to protect area residents from British invaders. They did not, however, keep the British from landing. During one episode, the invaders seized property from islanders Harry Keene, Valentine Barnes, William Barnes, and

William Geohegan. On another occasion, a British ship anchored off the mouth of what is today Punch Island Creek and put a boat over to go ashore. An islander by the name of Tall fired at them and, probably much to his surprise, watched the Redcoats return peacefully to their ship. They no doubt thought the militia lay in wait for them.

British vessels were again off Taylors Island in the winter of 1814, when the warship *Dauntless* lay at anchor in the Patuxent River across the bay. Her captain sent several tenders out to forage for supplies. One raid took them to the Little Choptank and the port of Tobacco

Stick, now Madison. They anchored for the night on the east side of James Island, and the crew awoke the next morning to find the vessel locked in ice and surrounded by a company of volunteers. Capt. James Stewart and his men, who had crossed the ice in the night, marched their captives to Easton. They also took along Becky, a slave woman whom the British had captured. In the meantime the tender was dismantled, and the spoils were auctioned off at Jeremiah Spicer's store. A cannon from the vessel eventually was mounted on the banks of Slaughter Creek on Taylors Island. The cannon was fired regularly on special oc-

Oak Grove Methodist Church on Golden Hill, built in the late nineteenth century, as it appeared in 1983. (Author photograph)

casions until it exploded during a celebration of Woodrow Wilson's presidential election. It is now mounted and appropriately marked in a little park at the west end of the bridge to the island.

From the early 1700s to 1850, Taylors Island forests provided an abundance of timber and a source of income for residents, who supplied lumber and lumber products to area shipyards. When the timber eventually ran out, many islanders supplemented their farm incomes with fishing. They began harvesting oysters in the 1830s, and that soon became the chief income-producing occupation on the island. It reached its peak during the "oyster boom" following the Civil War.

A growing population led to the establishment of the first Taylors Island post office in 1850 and construction of a bridge across Slaughter Creek in 1854. The present bridge is the fourth at that site. The 1877 *Lake, Griffing, and Stevenson Atlas of Dorchester County* shows the Griffith windmill at the head of Oyster Creek, the Kirwan windmill on St. John's Creek, and the Travers windmill on Hoopers Point on the Little Choptank River. No traces of these windmills remain.

Island shipyards built pungies, schooners, bugeyes, and sloops for oyster dredging. This and the lumber industry almost decimated the trees on the island. The last round-bottomed sloop, the *J. T. Leonard,* was built on Taylors Island in 1882 by Moses Geohegan. She worked the bay, dredging oysters and carrying freight until 1965. She was then moved to the Chesapeake Bay Maritime Museum at St. Michaels, where she remained for ten years until nothing could be done to stop her deterioration. Over the winter of 1975, ice opened her seams, and the *Leonard* sank in her museum berth. Museum boatyard workers refloated her, only to find that she was be-

yond saving. Before breaking up the sloop, they removed everything salvageable to create a museum exhibit.

Unlike the *Leonard,* many of the island's houses have survived. Others besides Mulberry Grove are still used. Among them is Oyster Creek Farm (ca.1830), which is across Oyster Cove from Mulberry Grove. The farm boasts a five-hole frame privy that came from Mulberry Grove. Overlooking Oyster Creek is Patrick's Discovery (ca.1870), a well-built example of Dorchester County architecture. The island churches show the deep religious background often in evidence on the Eastern Shore. In 1781 islanders formed the Methodist Society, and they built the Taylors Island Meeting House in 1787. The congregation replaced the meeting house in 1857 with the Bethlehem Methodist Church, now known as the "Old Brick Church." Before the Civil War, the congregation split over the issue of slavery. Other churches include the Lane Methodist Episcopal and the Chaplin Memorial Church, both built in the late nineteenth century.

A few islanders continue to farm, but marine activities are their main source of income. Commercial fishing, primarily for fish, oysters, crabs, and soft-shell clams, has declined, garnering near record low prices. In the Little Choptank River and Taylors Island area, the 1997 fish catch was valued at $18,000 for 76,484 pounds; 1,078,650 pounds of crabs brought $1,028,197, and 332,000 bushels of oysters sold for $625,979. The island's seafood catch in 2000 totaled 8,349 pounds of fish, 159 bushels of oysters, and 131,404 pounds of soft- and hard-shell crabs. In addition to the traditional water-related livelihoods, businesses catering to pleasure boaters and hunters are a mainstay of the island economy. The lower end of the main island is a 976-acre wildlife management area that provides opportunities for hunting waterfowl and Sitka and white-tailed deer.

## REFERENCES

1847 TC, USCGS, 250; 1877 Lake et al., *Atlas;* 1901 TC, USCGS, 2560; 1942 TC, USCGS, 8110; 1942 Taylors Island Q, USGS; 1974 Taylors Island Q, USGS (8,033.3 acres).

*Guide to Public Hunting Areas in Maryland,* MdDNR, Wildlife Administration (1979). *Guide to Public Piers and Boat Ramps,* MdDNR, Tidewater Administration (1982). Hayes, Anne M., and Harriet R. Hazleton, *Chesapeake Kaleidoscope* (Cambridge, Md.: Tidewater, 1975). Huelle, Walter E., *Footnotes to Dorchester History* (Centreville, Md: Tidewater, 1984). Jones, Elias, *New Revised History of Dorchester County* (Cambridge, Md.: Tidewater, 1966). *Maryland's Waterfront Facilities,* MdDNR, Tidewater Administration (n.d.). Mowbray, Calvin W., *Dorchester Tercentenary Bay Country Festival, 1669–1969,* Souvenir Book (1969). Mowbray, Calvin W., *Early Settlers of Dorchester County and Their Lands,* 2 vols. (private printing, 1981). Stump, Brice N., *A Visit with the Past* (Chicago: Aglams Press, 1968). Weeks, Christopher, ed., *Between the Nanticoke and the Choptank: An Architectural History of Dorchester County* (Baltimore: Johns Hopkins University Press, 1984).

# HOOPERS ISLAND
## *A True Waterman's Community*

Hoopers Island lies south of Taylors Island. About twenty miles long, it is actually three islands identified as Upper, Middle, and Lower Hoopers. A bridge connects Upper Island to the mainland at Meekins Neck, and another joins Middle and Upper Islands. Access to Lower Hoopers Island, which is mostly marsh, is by boat. The three islands separate the Honga River to the east from Tar Bay on the west and the Chesapeake beyond.

Most of the residents occupy three small villages, Honga, Fishing Creek, and Hoopersville. The Middle Island village of Hoopersville is the largest, with general stores and packing plants. To reach the packing house on Muddy Hook Cove, watermen must avoid two charted and submerged wrecks if they navigate the channel at high tide.

Indians were the first to use Hoopers Island, but accounts differ as to whether they were Choptanks of the Eastern Shore or Yaocomicoes from the other side. According to local legend, the area's first English settler was Henry

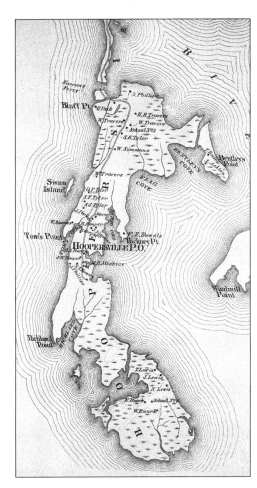

Hoopers Island, 1877, a prominent feature (and single island) in Lake, Griffing, and Stevenson's *Atlas of Dorchester County.* (Courtesy of Anne Arundel County Public Library)

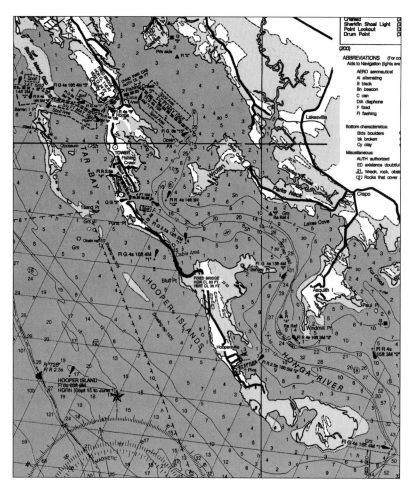

In 1848, the island occupied 4,190 acres. One hundred and fifty years later it had eroded to less than 550 acres, losing over 24 acres per year—the greatest erosion rate in the entire bay. Only one-eighth of this island remains today. (NOAA 12260, 2002. Courtesy of Maptech, Inc.)

and an officer in the militia between 1723 and 1744. Hooper died on April 20, 1767. That same day his house burned to the ground, and his body "being then in the house, was with much difficulty saved from the flames." At his death, he owned at least 2,900 acres in Dorchester County.

Henry Hooper Jr., only surviving son of the second Henry Hooper, was a planter and owned some 4,400 acres in Dorchester County. Henry Jr. served in the lower house of the legislature from 1768 to 1780, was a member of the Council of Safety for the Eastern Shore in 1775, a member of the militia, and a judge of the general court. Henry Jr.'s son William, a physician, continued the Hooper tradition, representing Dorchester County in the legislature between 1785 and 1787. A member of the Maryland Militia, William rose to the rank of captain of the Light Dragoons.

Although the Hoopers were the largest landowners, other early settlers also claimed land on the islands. William Chaplin patented 300 acres, which he called "Chaplin's Holme," and Richard Bentley patented another 300 acres on what is still known as Bentley Point. A third man, Philip Shapeley, patented 117 acres. In 1670, when they earned their freedom, several indentured servants acquired fifty-acre parcels, along with clothing and farming supplies. One was Thomas Hooten of "Swan Isle." Daniel Puddiford of "Puddiford's Chance," received fifty acres in 1672.

Where the bridge crosses Fishing Creek, the Maryland assembly established a town in 1683. This was the site of Plymouth, a town created for the "advancement of trade." Only eight of the proposed one hundred lots were taken up, and growth of the town slowed. In 1748, a tobacco warehouse

Hooper, who came across the bay from Calvert County around 1667. Married twice, Hooper had fourteen children, and in time the clan became one of the largest on the shore. Henry Hooper was a planter, served as a justice for Dorchester County, and represented the county in the General Assembly in the 1790s. The island that bore his name was part of more than 2,400 acres that Hooper owned in the county at the time of his death in 1720, but, by all accounts, he never lived on it.

Henry Hooper's son, a second Henry, was a mariner, planter, and attorney. He followed his father into public service and became a legislator and speaker of the lower house. He was also a judge and a justice of the provincial court, a member of the school board,

became Plymouth's chief commercial draw and operated until 1773. The name Plymouth was last used in 1776 when the local militia company called themselves the Plymouth Greens. The site of Plymouth is now a county park.

During the Revolution, when British troops raided Hoopers Island for supplies, the legislature moved quickly, sending a small body of troops to defend the island from further incursions. Records show that Henry Hooper Jr., who was a member of the General As-

sembly at the time, appealed to his fellow lawmakers for money to purchase rations for the troops on Hoopers Island. After the Revolution, tobacco gave way to vegetables and grain as the islanders' cash crop. During the War of 1812, the British forces on the bay and Tory renegades were a constant threat to Hoopers Island residents, but no actual hostile engagement ever took place.

Ferry service from Fishing Bay through Hoopers Strait to Hoopers Island was established as early as 1786.

Motor vehicles on the waterfront of Lower Hoopers Island, 1930s. (The Maryland Historical Society, Baltimore, Maryland)

*Top:* Hoopers Island Memorial Methodist Church, erected in 1877. (Author photograph, 1983)

*Bottom:* The Hoopers Island lighthouse, built in 1902, was automated by the end of the century. (Courtesy of the Chesapeake Bay Maritime Museum, photograph by Jerry Land)

There had long been a Fishing Creek Ferry, and the *Lake, Griffing, and Stevenson Atlas of Talbot and Dorchester Counties* of 1877 shows a Narrows Ferry between Upper and Lower Islands. Until 1901, when a bridge was built, a scow served as a ferry between Upper and Middle Islands. Lost in the hurricane of 1933, the 1901 bridge was rebuilt two years later. It was replaced by the present high bridge, which was completed in 1979.

By the middle 1880s, islanders enjoyed steamboat service. A steamboat wharf at Hickory Cove was a regular stop until 1929. The trip to and from Baltimore cost five dollars, and another two dollars for a private stateroom. The boat left Hoopersville in the late afternoon on Monday, Wednesday, and Friday, and arrived in Baltimore at about 3:30 the following morning. It returned Tuesday, Thursday, and Saturday.

The Hoopers Island Lighthouse, which is actually off the island's shore, suffered several delays before the caisson was towed to the site in July 1901. It began operation in June 1902. Rising sixty-five feet above the water, the light is now fully automated and flashes every six seconds.

During the oyster boom in the mid-1880s, many immigrant Germans were shanghaied from Baltimore to provide much-needed crews for the oyster fleet. Often unable to speak English and unfamiliar with their surroundings, the crewmen were housed in so-called Paddy shacks, which were little more than prison compounds. The German Society of Baltimore hired a tugboat and turned it over to a U.S. marshal who raided and destroyed several Paddy shacks on Hoopers Island.

Lower Hoopers Island, which is no longer habitable, was once the thriving community of Applegarth, which grew from less than half a dozen families before the oyster boom. By 1893 the community of more than one hundred people had a post office, several stores, and an elementary school. Erosion was a serious problem by the turn of the twentieth century, and many residents were forced to leave. The post office closed in 1924, and the last people abandoned the island in the late 1920s. Farmers continued to pasture cattle and sheep there until the devastating 1933 hurricane washed out the bridge, which was never replaced.

Several of the oldest buildings on Middle Hoopers Island have survived. They include the Swan Island Hunting Club built in the early 1800s. Swan Island itself has almost disappeared, but the hunting club still stands on Middle Hoopers Island. Another landmark is the Parr House in Hoopersville. Typical of the inherent adaptability of vernacular architecture, the frame dwelling is built in five parts, each of which has a ground-floor room. The Hoopersville Community Hall and Meeting House, which was built originally as a chapel about 1885, is still used for social and political meetings. Founded in 1780, the large Island Memorial Church is a center of community life. During the Civil War, the differences that divided the nation divided the members of the church. Part of the congregation broke away to found the Methodist Episcopal Church South and built their own chapel.

Good roads, telephones, radio, and television brought the twentieth century to Hoopers Islanders, but they still are unique in their orientation to the water rather than the land. Residents are never any great distance from water. In places, houses on both sides of Route 335, which runs the length of Upper and Middle Islands, have waterfront, some on the bay and some on the Honga River. Originally farmers, almost all of the residents now make their living from the water as crabbers, oyster tongers, or seafood packers. Until recent years, when the health of the bay's crabs, oysters, and other seafood has periodically diminished their supplies, seafood proved a lucrative business. In 2000 Hoopers Island workers hauled in 1,109,708 pounds of fish, 1,795 bushels of oysters, and 1,447,034 soft and hard crabs.

In addition to being a commercial fishing center, Hoopers Island has been a sport-fishing mecca since 1972. Every June the volunteer fire department sponsors the island's largest event, the William T. Ruark Fishing Tournament, named in honor of a popular charter boat captain and local notable. Organizers offer prizes for the biggest bluefish, trout, hardhead, flounder, and drumfish. Fishing and the natural attractions of Hoopers Island play a major role in the island economy, which is supplemented by other popular local events, including an Arts and Crafts Auction and Bazaar in October and, in November, the Ronald McGlaughlin Artisans Fair.

The long profile the three islands turn to the bay has made them vulnerable to erosion, which has averaged 24.19 acres per year. The most dramatic land loss has occurred on Lower Hoopers Island, which has turned into marshland and a habitat for waterfowl. Dorchester County boasts the largest marshland acreage in Maryland thanks, in part, to Lower Hoopers Island. Along its entire length, Hoopers—Upper, Middle, and Lower—is a haven for a variety of waterfowl. Years ago, such large flocks flew over the island that homeowners had to put bars across their windows to protect them from the birds. Today, diving ducks congregate on the waters surrounding the island, and ranging its

shores are little and great blue herons; American oyster catchers; double-crested cormorants; willets; herring gulls; Forster's, common, royal, and least terns; black ducks; gadwalls; and boat-tailed grackles.

REFERENCES

1848 HC, USCGS, 209 (Upper 1,271.41 acres, Middle 2,119.66 acres, Lower 805.55 acres, total 4,196.62 acres); 1848 Richland Point Q, USGS; 1877 Lake et al., *Atlas;* 1905 Ewell Q, USGS; 1942 Ewell Q, USGS (Upper 1,031 acres, Middle 1,655 acres, Lower 657 acres, total 3,343 acres); 1973 Richland Point Q, USGS; 1974 PR Honga Q, USGS; 1984 NOAA 12261 (Upper 859.01 acres, Middle 1,702.73 acres, Lower 791.96 acres, total 3,353.70 acres); 1998 NOAA 12261 (Upper 357.11 acres, Middle 133.91 acres, Lower 76.52 acres, total 567.54 acres). Erosion loss 24.19 acres per year.
Arnett et al. Dilley, Ray, "The Distinctive Hooper Islander," *Bay Sailor,* June 4, 1984. Jones, Elias, *New Revised History of Dorchester County.* Meanley, Brooke, *Waterfowl of the Chesapeake Bay Country* (Centreville, Md.: Tidewater, 1982). Mowbray, Calvin W., *Early Dorchester* (private printing, 1979). Mowbray, Calvin W., and Mary I. Mowbray, *The Early Settlers of Dorchester County and Their Lands,* 2 vols. (private printing, 1981). Norden, A. W., O. C. Forester, and S. Fenwick, eds., *Threatened and Endangered Plants and Animals of Maryland,* Special Publication 84-I, Maryland Natural Heritage Program (Annapolis: MdDNR, 1984). Stinson, Anne, *Hoopers Island Today and Many Yesterdays* (Easton, Md.: Easton Publishing Company, 1975). Weeks, Christopher, ed. *Between the Nanticoke and the Choptank, An Architectural History of Dorchester County* (Baltimore: Johns Hopkins University Press and MdHT, 1984). Wennersten, J. R., *The Oyster Wars of Chesapeake Bay* (Centreville, Md.: Tidewater, 1981).

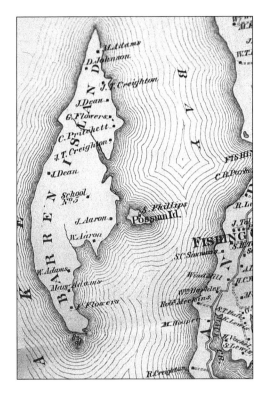

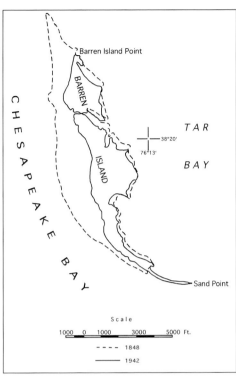

*Far left:* Barren and nearby Possum Islands, 1877. Barren, Lake, Griffing, and Stevenson's *Atlas of Dorchester County* showed that at the time eleven farms and a school dotted the island.
(Courtesy of Anne Arundel County Public Library)

*Left:* Erosion drawing of Barren Island, 1848 to 1942.
(Courtesy Maryland Board of Natural Resources, Bulletin #6, 1949)

Barren Island once flourished as a farm community with its own church, school, and a store or two. Now, true to its name, it lies barren—abandoned to the elements, migrating waterfowl, and memories. Gone are the days when "Barn" Island, as many Eastern Shore natives call it, consisted of more than 580 acres of fertile farmland and bountiful hunting grounds.

Before the English settlers arrived, Nanticoke Indians camped on the island's shores to hunt and fish. Recent visitors still find arrowheads and other Indian artifacts but no sign of permanent villages. The English discovered Barren Island in the mid-1600s. A string of shoals north and south of the island confirm that once it was a long peninsula. Today, to the north, a channel allows passage from the bay to the Honga River. Occasionally, when the tide ebbs

across Barren Island Gap, unwary sailors run their vessels aground against the down-tide side of the channel. Early Marylanders originally called this shallow waterway Preston Creek in honor of the first colonist to lay claim to the island.

Richard Preston arrived in Maryland from Virginia around 1650. With his wife and seven children, he was among a group of Puritans seeking greater religious freedom in the colony of Cecil Calvert, second Lord Baltimore. Settling first in Calvert County, Preston became a wealthy planter and an important man in early Maryland politics and government. In the mid-1650s he and other pro-Parliament, anti-Catholic, anti-Calvert commissioners briefly seized control of the government. After Cecil Calvert regained control, Preston continued to represent Calvert County in the lower house of the General Assembly. He also converted to

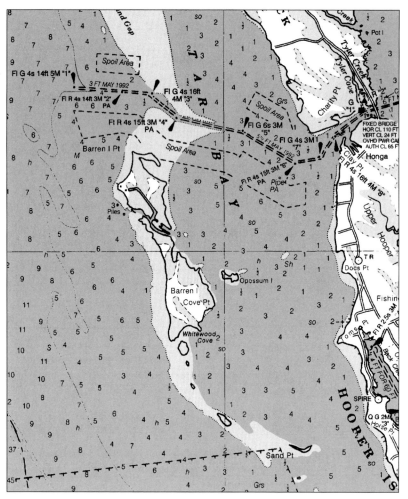

In the early twenty-first century, having eroded from 582 acres in 1848 to 116 in 1998 and completely abandoned, Barren Island supplied refuge to several large colonies of shorebirds.
(NOAA 12231, 2002. Courtesy of Maptech, Inc.)

Quakerism, prompting Charles Calvert, governor of Maryland in the 1660s, to refer to him as the "Great Quaker." Apparently overlooking past political differences, Lord Baltimore granted Richard Preston the right to the Dorchester County island in 1664. Preston paid Calvert a quit rent of fourteen shillings a year, with "Royall mines excepted." The last was a provision added in case the colonists discovered the mineral riches that British monarchs hoped one day would fill their coffers. Preston eventually moved to Dorchester County and was its first delegate to the General Assembly. He died around 1669, and the island passed into other hands.

Farming was the chief eighteenth- and nineteenth-century enterprise on

Barren Island and on little Opossum Island to the east. At the height of its productivity, Barren supported fourteen farms. Another was on what people then called Possum Island. Erosion has since reduced Opossum Island to an islet only visible at low tide. The residents of Barren Island worshiped in a Methodist Church served by a visiting parson. They had their own stores to supply basic needs and a one-room school for their children.

Emigration began in the early 1900s as erosion ate away at the island, making its residents increasingly vulnerable to storms and high tides. Water spread over fields, flooded roadways, and eventually lapped at building foundations. One by one, farmers loaded their possessions on barges and moved to Hoopers Island or the mainland across Tar Bay. The last family left in 1916.

Like the Indians hundreds of years before, men continued to use Barren Island for hunting. A group of wildfowl hunters built a lodge on the Chesapeake side in 1929. They used materials salvaged from the demolition of the Caswell Hotel in Baltimore and transported by boat via Solomons Island across the bay.

Some years later, Baltimore lawyer William Siskin bought Barren Island and the lodge. Among the friends and political allies Siskin entertained there in the late 1970s and early 1980s was former governor Marvin Mandel. By that time, however, erosion threatened the bulkheading along the bay shore. Siskin's pleas for help from Dorchester County and the U.S. Army Corps of Engineers went unheeded. He failed to persuade them that his island served as a barrier protecting Hoopers Island to the east and warranted government action to save it. The bay continued its work, relentlessly breaking down bulkheading

*Top:* The inviting and fully functional Barren Island Clubhouse, circa 1950.
(Courtesy of the Chesapeake Bay Maritime Museum)

*Bottom:* The Barren Island Clubhouse brought to grief, 1987. When sportsmen neglected the bulkhead protecting the building, water soon reached the clubhouse, completely washing it away by 2002. Only the foundation, now under water, survives.
(Author photograph)

and leaving the lodge defenseless. Abandoned to the elements, the bulkhead finally collapsed in 1985, and waters of the bay were soon lapping around the building itself. Today, the lodge is gone, only its foundation is visible under water.

Man has not been the only occupant of the island. Once, least terns flourished all along the coastal waterways from Massachusetts to Florida. Arriving in the Chesapeake in April, they nested on Barren Island's isolated beaches, making a simple scrape in the sand above the high water mark. By August their young were ready to join them as they took to the sky, beginning their fall migration to

South America. Man ended this state of affairs when the birds fell prey to Victorian fashion and a growing millinery industry insatiable for the tern's handsome plumage. In the late 1800s, their feathers sold for ten to twelve cents apiece. Crafty hunters lured the curious birds to their deaths by tying a white rag to a stick and throwing it into the air. In a few moments a group of terns swooped down, mistaking the rag for a fallen comrade. Within ten years professional hunters had decimated the population. The federal government finally stepped in and saved the graceful, gull-like birds from extinction. In 1913 it put them under full protection of the law. Beginning in the 1920s and on into the 1950s, the terns made a dramatic comeback on Barren Island. Soon, however, they fell victim to relentless erosion and beach development. One by one, their nesting places disappeared, and the birds left in search of new habitats. Surveyors in the 1980s recorded only three colonies of least terns in Maryland. By 1996, however, the number of colonies had risen to fifteen with a total of 550 breeding pairs, mostly on Assateague Island. Of the fifteen colonies, ten are nesting on rooftops, reflecting the loss of natural habitat.

In 1981 the U.S. Army Corps of Engineers dredged the Barren Island Gap Channel between the Chesapeake and Tar Bays. Grateful fishermen no longer had to travel the full length of Upper and Lower Hooper Islands to reach the Honga River. The dredging created a large spoil island off the northern end of Barren. There, the Maryland Department of Natural Resources created and maintained a bird refuge. To attract herons, gulls, and terns, DNR workers planted marsh grass along the island's shore.

At the same time, a Baltimore attorney, Raymond Simmons Jr., claimed the manmade island, maintaining that no one owned it. He even went as far as to name it Steve's Island after his son. The state contested his right to the island and took him to court. After two years the court finally decided in favor of the Department of Natural Resources. Moving ahead with its original plan, the DNR extended the island and created the Tar Bay Wildlife Management Area in 1985.

Over the years since, several people have had designs on Barren Island, which in 1988 was valued at $250,000. One of the rejected proposals was a $25 million, 200-slip marina with lodges and cabins to be built on the northern end of the island. Another was for a new state prison—a sort of "Alcatraz by the Bay." The state was not interested, and the buyer backed out of the deal. Subsequently, the Department of Natural Resources raised its price to $495,000, discouraging other prospective developers. Increased impediments to navigation have also decreased the island's attractiveness as an investment. Despite recent dredging, silting has filled the channel from a dredged depth of seven feet to about four feet, making it risky to navigate for those unfamiliar with the waters.

The U.S. Fish and Wildlife Service now owns Barren Island. To halt erosion, the agency installed a series of geotubes, or protective shoreline barriers, filled with material dredged from the Tar Bay Channel. Unfortunately, the material with which the original tubes were filled was fine-grained rather than coarse-grained sand. They collapsed, and the agency replaced them in the fall of 1998. Behind these barriers, the island is about 75 percent salt marsh. On the north, the land is low, flat, and covered with scrub growth. The southern end, however, offers higher ground with pine stands, meadows, and wetlands. Island pines accommodate a large heron rookery of 350 nests, and wildlife observers have spotted at least one eagle's nest. Shallow waters surrounding the island serve as a prime winter habitat for large concentrations of red-head, canvasback, and black ducks. Today the island may be bereft of human beings, but its feathered population makes it anything but barren.

## REFERENCES

1848 TC, USCGS, 255 (582.72 acres; Possum Island, 12.03 acres); Lake et al., *Atlas;* 1901 TC, USCGS, 2564; 1932 TC, USCGS, 4710; 1942 Barren Island Q, USGS (368.19 acres); 1971 MdDNR-11R6 49–51; 1974 PR Honga Q, USGS; 1983 NOAA 12264 (254.9 acres); 1984 NOAA 12261 (two large, five small islands, 206.86 acres total; Possum Island 5.10 acres); 1998 NOAA 12261 (one large, three small islands, 116.21 acres total; Possum Island, tidal only). Average erosion loss approximately five acres per year.

Brunori, Carlo, MdDNR, personal communication with author, October 14, 1984. Earhart, H. Glenn, and E. W. Garbisch Jr., "Habitat Development: Utilizing Dredged Material at Barren Island, Dorchester County, Maryland," *Wetlands* 3 (1983). "Hunters Claim Island State Says is for the Birds," *Sun,* September 13, 1983. Meneely, Jane, "New Life for Terns," *CBMag* 13 (August 1983). Mowbray. "Nobody Wants to Buy Island off Maryland," *The Record* (Havre de Grace, Md.), December 3, 1981. Singewald, J. T., Jr., and T. H. Slaughter, "Shore Erosion in Tidewater Maryland," MdGS, *Bulletin* 6 (1949).

## BLOODSWORTH ISLAND
*A Tale of Erosion's Relentless Toll*

Bordered by Hooper Strait to the north, Holland Straits to the south, Deal Island to the east, and the Chesapeake Bay to the west, Bloodsworth Island's history of occupation stretches back to 1672. That year George Thompson of St. Mary's County on the western side of the bay acquired the island. He never lived there, however, and Thompson's Island passed through a number of hands over the ensuing years. The names Bowles, Woolford, and Hopkins appear on deeds tracing ownership into the 1750s.

Finally in 1759, Comfort and William Hopkins sold the island to attorney Charles Goldsborough, a wealthy landowner who represented Dorchester County in Maryland's General Assembly. Identified thereafter as Goldsborough Island, it passed to Charles's son Robert, and when Robert died, it went to his youngest son Howes. The entire island was sold to Robert Bloodsworth in 1799, from which time it was the property of resident owners who made their living from the land and the surrounding waters.

Because very little of the island was arable, it was a marginal area. In the early nineteenth century, when only about thirty of the island's more than five thousand acres would support farming, a few residents—mostly Robert Bloodsworth's descendants—eked out a living. They grew food for themselves and their livestock but never had enough tillable land to produce surplus crops for sale. High tides, sinking land, salt water intrusion, and relentless erosion ended their efforts to farm. By 1850 they had turned to the water and between 1860 and 1880, worked mainly as oystermen.

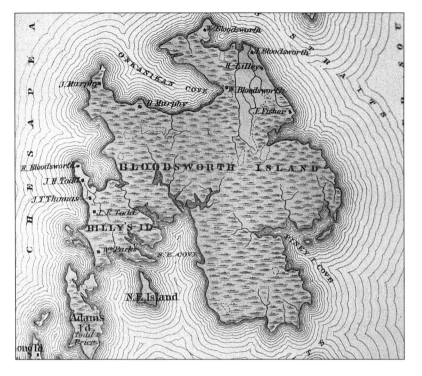

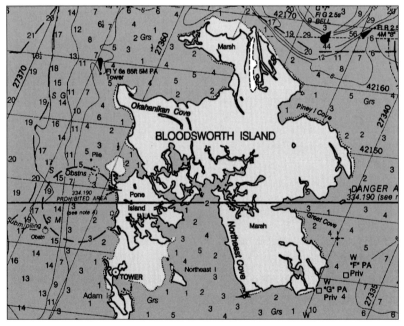

*Top:* Bloodsworth Island, 1877, as depicted in Lake, Griffing, and Stevenson's *Atlas of Dorchester County.* Nine farms occupied the high grounds, and swampland formed nearly all the rest of the island. (Courtesy of Anne Arundel County Public Library)

*Bottom:* Since 1945 a U.S. Navy practice bombing and amphibious landing site, Bloodsworth has eroded from 5,690 acres in 1849 to 4,208 in 2002, an average loss of almost 10 acres per year. (NOAA 12231, 2002. Courtesy of Maptech, Inc.)

County assessments suggest that even as watermen, the islanders were marginal operators. In 1876, for instance, John Bloodsworth paid taxes on ten acres of arable ground worth $125, plus $25 in improvements; a large sail canoe valued at $50; two yoke of oxen; and nine cattle, eight sheep, and ten hogs. Twenty years later his tax assessment tells the story of the severe erosion taking place on the island. By 1896 John Bloodsworth's land was listed as a single arable acre and the rest marshland valued at $11 with a dwelling worth $40, no livestock, and only one very small canoe.

Bloodsworth's house was typical of island buildings, which also tended to reflect the marginal character of the place. Whereas farmhouses on the mainland and on nearby Hollands Island were assessed at $200 to $600, most of those on Bloodsworth Island ranged in value from $25 to $100. The most valuable house was that of Job Murphy, assessed at $150 in 1876.

As more and more of the island became salt marsh, residents moved to the mainland, selling their land to outsiders for hunting or for oyster and crabbing operations. By the 1920s a gunning club, the Bloodsworth Island Game Company, owned the lower part of the island, and twenty years later the Phillips Packing Company bought the upper island and the nearby oyster grounds. The Game Company sold its section to the U.S. government for a navy bombing range in 1945, and by 1948 the entire island was the property of the navy. It is still used extensively for a bombing and ordnance testing range and for training navy Seal teams.

The significant ecological changes in the character of the island and the resulting depletion of upland area caused the loss of much of Bloodsworth Island's wildlife. Deer, rabbits, and other small animals, except for muskrats and raccoons, disappeared along with most trees and bushes. Vegetation changed to more salt-tolerant species such as marsh elder, groundsel trees, cordgrass, and salt grass, with black needle rush predominating. Several great blue heron rookeries still exist, along with a few great egret and yellow-crowned night heron nesting areas. Ospreys flourish. During their seasonal migrations, many species of ducks, geese, swans, shorebirds, and songbirds stop over on the island.

Land loss can be expected to continue at existing or higher rates in the coming decades. Given the deteriorating interior conditions of the marsh, another rapid rise in sea level could cause an even greater loss of land and possibly the disappearance of the island entirely.

Bloodsworth Island is closed to the public except by special permission from the navy.

REFERENCES

1849 TC, USCGS, 269 (5,687.9 acres); Lake et al., *Atlas* (5,561.6 acres); 1901 TC, USCGS, 2558; 1903 Bloodsworth Island Q, USGS ; 1942 TC, USCGS, 8134; 1973 Bloodsworth Island Q, USGS (4,950 acres); 1983 NOAA 12231(two large, six small, 4,396.77 acres); 1998 NOAA 12231. Average erosion loss 8.66 acres per year.

Downs, Lynda, Robert J. Nichols, Stephen Leatherman, and Joseph Houtzenroder, "Historic Evolution of a Marsh Island: Bloodsworth Island, Maryland," *Journal of Coastal Research* 10 (Fall 1994): 1031–44. McKewan, Sean, and Carlo Brunori, *Environmental Assessment for Bloodsworth Island Shore Bombardment* (Annapolis: Information Technology Service, Maryland Wildlife Administration, 1981). Mowbray. "Navy Seals Poised for Invasion: Bloodsworth Island in Dorchester County is the Target for the Elite Warfare Teams' Practices," *Sun*, April 25, 1997.

## HOLLAND ISLAND

*From Prosperous Town to Hunting Preserve*

The southernmost land in Dorchester County, Holland Island is two miles southwest of Bloodsworth Island, with the open bay to its west and Holland Straits to the east. It is now deserted. The island's name can be traced back to a seventeenth-century colonist, Daniel Holland, who bought the property from Thomas Taylor, sheriff of Dorchester County.

During the Revolution, the waters of Holland Straits were the scene of considerable action as Maryland's navy attempted to curb British and loyalist raids on shipping, islands, and communities along the shores of the bay. Spring Island, just east of Holland Island, may have been the site of temporary British fortifications.

At the time of the Revolution, a small, low island, known as Long Island, lay directly north of Holland Island, and in 1795 both belonged to John Price. Price descendants were still living on Long Island in 1880, but by 1900 it had all but disappeared, and the Prices had moved to Holland Island. One notable family member was Ephraim Price, who came to be known as the "father of Methodism on Holland's Island."

Half a dozen families made up the first Holland Island community in the 1850s. Over the next thirty years, the population grew to more than 360 people and more than seventy homes, stores, and other buildings. Most of the men worked the water, dredging oysters in winter and crabbing and fishing in the

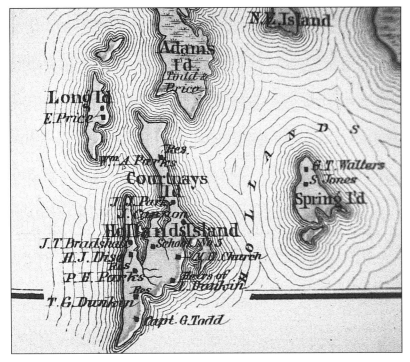

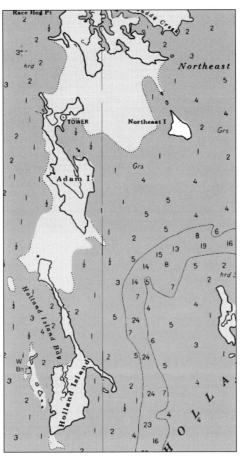

*Top:* Holland Island, 1877, as it appeared in Lake, Griffing, and Stevenson's *Atlas of Dorchester County.* (Courtesy of Anne Arundel County Public Library)

*Bottom:* Holland Island has lost 167 acres in the past century and a half, averaging a loss of slightly more than 1 acre a year. Just 21 acres remained in 2002. (NOAA 12231, 2002. Courtesy of Maptech, Inc.)

summer. A smaller number farmed, growing fruit, vegetables, wheat, and corn. Almost every family had chickens and ducks, and the eggs they didn't use were shipped to market in Baltimore.

In 1888 residents built the Hopkins Chapel Methodist Episcopal Church on the lower part of the island with George F. Hopkins as its first preacher. Later renamed the Holland Island Methodist Episcopal Church, the building could seat the island's entire population of 367 people.

The federal government authorized the building of the Holland Island Bar Lighthouse off the southern tip in 1889. It guided vessels safely into the Kedges Strait to the south. The screwpile lighthouse was a hexagonal frame building that served until 1960, when it was dismantled and replaced by an automated beacon. The lighthouse was the scene of several unusual events, including the mysterious death of the keeper in 1931. Even more bizarre was the 1957 attack by three navy fighter bombers from the Atlantic City Naval Air Station. They mistook the lighthouse for a nearby sunken ship that was their intended target and fired seven rockets, three of which hit the lighthouse. The warheads did not contain live explosives but still did considerable damage. The four lighthouse keepers, who escaped serious injuries, contacted the Coast Guard to alert the Patuxent Naval Base across the bay. The navy sent a doctor, who arrived by helicopter and was lowered in a sling to give first aid to the two keepers who were injured. A Coast Guard vessel evacuated the keepers, but they returned the next day and began repairing the damage. The present Holland Island Bar light is a quick flashing light thirty-seven feet above the water.

In 1914 Holland Island was about five miles long and one and a half miles

wide. It had a post office, a church, a two-room schoolhouse with two teachers, several stores, and a community hall for holding oyster suppers and box socials. At one point it even had its own doctor and a championship baseball team. Two African-American families lived on the island. The men built boats and dredged for oysters. One of the women, Mary Rogers, was a midwife who delivered many of the children born during her time there.

The islanders were self-sufficient and well prepared for the times in winter when ice made it impossible for anyone to get onto or away from the island. They kept large quantities of coal and wood on hand, salted many barrels of fish, and raised numerous hogs, geese, ducks, and chickens. In the winter, the surrounding waters were filled with wild ducks and geese, and the hunting season usually yielded enough game to sell on the mainland.

As time passed, more and more men went to work on the water. Irving Parks recalled that the work boat fleet consisted of eight pungies, thirty-six bugeyes, forty-one skipjacks, and two schooners, some of which had been built on the island. Besides dredging for oysters, Holland Islanders used their boats for crabbing and setting pound nets to catch shad west of the island.

In 1914 the first signs of serious erosion of the high ground along the Bay Shore Ridge marked the beginning of the islanders' exodus to the mainland. Some simply moved to other parts of the island, but over the next four years, growing numbers left, taking houses, sheds, and sometimes their picket fences with them. Everything that could be saved was loaded on boats and sailed to Crisfield,

Eastern Ridge—the Main Street (or its equivalent) of Holland Island, late in the nineteenth century. Residents eventually moved these houses en masse to Cambridge. (Courtesy of the Chesapeake Bay Maritime Museum)

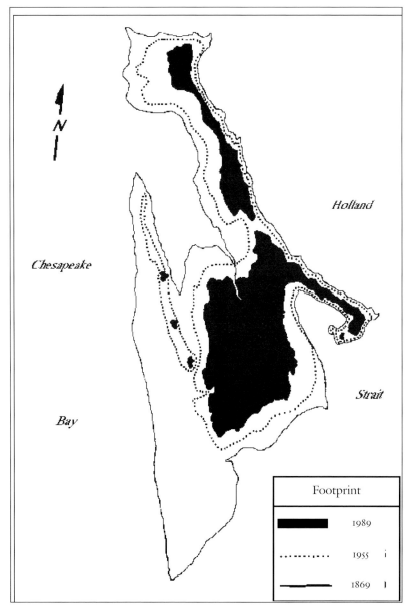

his parent's house, which is the last remaining on the island.

Since 1912, erosion has reduced Holland Island to 121.36 acres, most of which are marshland. A prime haven for nesting herons, the island supports a large rookery that numbered 609 nesting pairs in a 1995 survey. Gulls and terns nest on Holland Island, which is also a habitat for foxes, raccoons, mink, and even a small number of goats.

A dense growth of small persimmon trees, trumpet creeper, poison ivy, red cedar, and a few large deciduous trees and red cedar cover the northern section of the island. Farther down, yellow-crowned night herons and great egrets have claimed a parklike area grown over with large American hackberries. A long, relatively clear section leads to the hunting lodge, a few crumbling building foundations, and the Parks family graveyard. Trees around the place include large winged sumac, silver poplars, and a pear.

North and South Bars are sandy areas and are usually occupied by roosting gulls, terns, and other shorebirds. Between Southeast Hammock and East Central Hammock, on a high area in the marsh, is another graveyard with about fifteen poison ivy–covered graves and stones marked with the names of Prices and Dizes. Most of the island's salt marsh lies in the area of the southern hammocks. Loss of habitat has cut down the number of usual predators, and the island has become something of a sanctuary for songbirds and other native birds, including wrens, swallows, catbirds, blackbirds, song sparrows, fish crows, and grackles. It is truly a bird watcher's paradise.

A Salisbury resident and the island's latest owner, Stephen L. White, recently created the Holland Island Preservation Foundation with the hope of eventually shoring up what remains of this historic

*Above:* Erosion drawing shows change in island size and shape from 1869 to 1989, with complete loss of the "hammock" on the bay side where most of the houses were located. (Courtesy of Stephen L. White, Holland Island Preservation Foundation)

*Previous pages:* Aubrey Bodine's photograph of the last remaining house on Holland Island in 1953. (Bodine Collection, The Maryland Historical Society, Baltimore, Maryland)

Cambridge, Nanticoke, and other places. The schoolhouse was moved to Crisfield where it became a warehouse. When a tropical storm hit the bay in August 1918, it nearly destroyed the island church, moving it seven inches on its foundation, and that may have been the event that caused the last families to leave. The church building was moved to Fairmount, and by 1922 everyone had gone. Some returned for the summer crabbing season for a few years. Irving Parks did so for seventeen summers. A gun club used

island and perhaps stopping it from going the way of so many vanished islands of the Chesapeake Bay.

REFERENCES

1848 HC, USCGS, 211 (Holland and Courtenays islands 288 acres); 1877 Lake et al., *Atlas;* 1901 TC, USCGS, 2558; 1903 Bloodsworth Q, USGS; 1912 HC, USCGS, 3379 (192.82 acres); 1942 Bloodsworth Q, USGS: 1942 HC, USCGS, 6679 (157.01 acres); 1971 MdDNR DO I-I 3RL-22 (144.18 acres); 1972 Kedges Straits Q, USGS; 1973 Bloodsworth Q, USGS; 1984 NOAA 12228 (142.84 acres); 1998 NOAA 12228 (121.36 acres). Erosion loss 1.11 acres per year.
Arenstam, Sheila Jane, "Sea Level Rise and Environmental Refugees: A Case Study of the Abandonment of Holland Island" (master's thesis, University of Maryland, 1994). Armistead, Henry T., "Summer Birds of Chesapeake Bay Islands," *Maryland Birdlife* 34, no. 31 (September 1978). Dunkle, Maurice A., *Holland Island* (private printing, 1985). Hayes, Anne M., and Harriet R. Hazleton, *Chesapeake Kaleidoscope* (Cambridge, Md.: Tidewater, 1975). "Hollands Island: Residents Pack up Lock Stock and House," *Cambridge Daily Banner,* August 28, 1985. Hornberger/Turbeyville. Jones, Elias, *New Revised History of Dorchester County* (Cambridge, Md.: Tidewater, 1966). McAlister, James A., *Abstracts from Land Records of Dorchester County* (private printing, 1960). Mowbray. Parks, Irving M., Sr., *The Vanishing Island: The Holland Island Saga* (private printing, 1972). Parks, Mrs. William W., "I Remember When the Bay Forced a Village to Move," *Sun Mag,* October 18, 1953. Travillian, R., and F. Carter, *Treasure on the Chesapeake Bay* (Glen Burnie, Md.: Spyglass Enterprises, 1983). Wilson, Johanna, "The Death of Holland's Island" (valedictory essay, Crisfield High School, 1970).

## GREAT AND LITTLE FOX ISLANDS
### *From Famous Gunning Club to Education Center*

Great and Little Fox Islands actually represent an island group totaling some sixty-seven acres, most of which is marshland. In 1895 the group covered 357 acres and included seven islands: Great Fox, Clump, Northeast and South Little Fox, House, and North and South islets. These Virginia islands lie six miles south of Crisfield and are the tip of a peninsula that once divided Tangier and Pocomoke Sounds.

Cartographer Augustine Herrman recorded their presence in 1670 and called them the "Bedor Islands" for reasons unknown. The first English settler to claim one of the group was a Thomas Welbourne, who patented eighty-three acres on Little Fox Island in 1678. Thereafter, both Little and Great Fox Islands passed through several owners.

During the Revolution, Virginia loyalists led by Capt. John Kidd preyed on shipping from a base on the islands.

After the war, he continued to be a threat. In November 1782 the Maryland navy's fleet of four sailing barges headed out of their Onancock anchorage to track down the Tory raiders. In trying to locate them, the Maryland State Navy commodore Zedekiah Walley sent Capt. Levin Handy and Lt. Samuel Handy ashore on Fox Island in search of information on the whereabouts of the Tory fleet. Their informants sent them on to Cager's or Kedges Strait, where Kidd's superior force and the Marylanders' bad luck resulted in a loyalist victory at the bloody Battle of Cager's Strait.

The nineteenth century passed without incident in the life of the islands. Their history is notable mainly for the steady encroachment of Tangier Sound, creating as many as seven islands in the 1890s. As did many bay islands, Great and Little Fox attracted migrating ducks

*Right:* Great Fox and Little Fox Islands, 1849. (Courtesy of National Archives (NACP) RG 23, T272)

*Far right:* At the beginning of the twenty-first century, the Chesapeake Bay Foundation held title to what remains of the islands, maintaining them as a training school for conservation-related activities. The two islands, 589 acres in 1849, eroded away to form seven small islands that totaled 90 acres in 1998. They average a loss of 3.4 acres per year. (NOAA 12228, 2002. Courtesy of Maptech, Inc.)

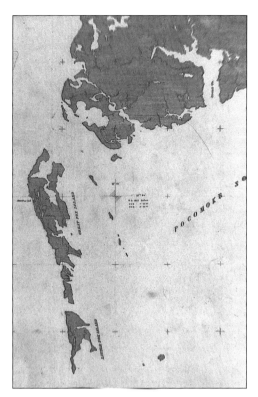

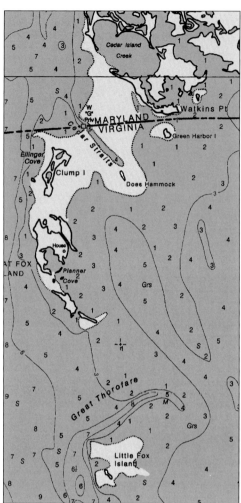

and geese in the fall and winter. In 1900 a group of hunters built a lodge on Great Fox Island and rebuilt it when it burned in 1920. Although using corn as bait to attract ducks and geese was illegal, the hunters kept up the practice. They kept their bait corn in a wooden shed at the end of their dock, and in 1974 a game warden discovered it. When the miscreants were brought to trial, the judge gave them the choice of donating the island to a worthwhile organization or going to jail. Not surprisingly, they chose the former option and donated Great Fox Island to the Chesapeake Bay Foundation in 1975.

The Chesapeake Bay Foundation operates the Fox Island Study Center, with facilities that include ten double-occupancy bedrooms, a meeting room, research vessels, and canoes. Its instructors lead groups of middle- and high-school students and adults on teaching tours through one of the most remote and biologically productive sections of the bay.

## REFERENCES

1670 Augustine Herrman Map, 1685 Christopher Browne Map, Papenfuse/Coale; 1849–52 TC, USCGS, 272 (Great Fox 474.84 acres, Little Fox 114.55 acres, total 589.39 acres); 1869 HC, USCGS, 997 (Great Fox 364.63 acres, Little Fox 128.09 acres, total 492.72 acres); 1895 HC, USCGS, 2397 (House Island 13.95 acres, Clump Island 36.87 acres, Great Fox 229.75 acres, North islet 6.32 acres, South islet 6.90 acres, N. Little Fox 27.91 acres, S. Little Fox 35.40 acres, total 357.10 acres); 1900 Crisfield Q, USGS; 1942 Ewell Q, USGS; 1942 Tangier Q, USGS; 1942 TC, USCGS, 8162; 1951 HC, USCGS, 7944; 1962 Tangier Q, USGS; 1967 Ewell Q USGS; 1968 Great Fox Q, USGS (Great Fox, three islands, 67.01 acres total; Clump Island 37.63 acres, South islet 14.68 acres, Tidal Flats 0.91 acres, total 120.23 acres);

1973 Great Fox Q, USGS; 1984 NOAA 12228 (Great Fox 48.46 acres, Clump Island 20.40 acres, South islet 10.20 acres, Little Fox 2.55 acres, total 81.61 acres); 1998 NOAA 12228 (Great Fox 42.43 acres, Clump Island 23.34 acres, South islet 11.12 acres, NE islet 8.82 acres, three islets 4.24 acres, total 89.85 acres). Erosion loss 3.35 acres per year.

Chesapeake Bay Foundation, *Annual Report*, Annapolis, 1995. Eller, Ernest McNeill, ed., *Chesapeake Bay in the American Revolution*, Maryland Bicentennial Bookshelf (Centreville, Md.: Tidewater, 1981). Footner, Hulbert, *Rivers of the Eastern Shore* (Centreville, Md.: Tidewater, 1964). Horton, Tom, "Islands of the Chesapeake," *Sun*, July 11, 1982. Papenfuse et al. Ukens, Carol, "Foxes in the Bay," *CBMag* (January 1983). Wilson, Woodrow T., *Crisfield, Maryland, 1765–1976* (Baltimore: Gateway Press, 1976).

## WATTS ISLAND
*Pirate's Lair to Wildlife Refuge*

Lying south of Crisfield and three miles east of Tangier Island in the open waters where Tangier joins Pocomoke Sound, Watts Island was abandoned to herons, goats, and other wildlife in the 1940s when its last human resident died. To date, the bay has taken nearly 300 acres of the island, leaving just over 116 acres of wetlands that support little more than scrub bushes, poison ivy, chiggers, and wildfowl.

Before Capt. John Smith discovered the island in 1607, Indians hunted there and left hundreds of artifacts including arrow- and spearheads and pottery shards. The island's recorded history began with Nickolas Waddelow, who claimed 400 acres in 1652 and named his grant St. Gabriell's Island. By 1659 Waddelow had sold three-fourths of his acreage to Robert King, Gilbert Henderson, Robert Blake, and John Watts, the settler whose name would be forever linked to the island. In 1670 Augustine Herrman included "Wat's Island" on his map of Maryland.

In the late seventeenth century the wooded bluffs and excellent harbor made Watts an ideal pirates' lair from which marauders could prey on ships sailing up and down the bay. The infamous local pirate Rodger Makeele operated from Watts Island in the 1680s. According to historian Donald Shomette, he formed a confederacy of pirates who terrorized the Lower Bay, "violently assaulting, plundering, and robbing" citizens of Maryland and Virginia.

In 1743 the island became the property of the first of several generations of Parkers. The 1800 census showed fifteen people living on Watts Island: Robert Parker, his family, and five employees. An iron fence once enclosed a small family cemetery on the east side of the island, which was gradually giving way to the waters of Pocomoke Sound. The erosion threatened the tombstones that marked the final resting place of three adults and a child. In 1966, before the cemetery disappeared into the sound, the island's owner, G. Randolph Kleinfelter, rescued and reestablished it with a memorial plaque at Port Isabel, his home on nearby Tangier Island.

By the time the Parkers were laid to rest, they had lived through the trials of three wars. Although the war for American independence ended at Yorktown on October 19, 1781, Maryland and Virginia loyalists were not ready to give up. Their vessels continued to harass and plunder the property of citizens along the lower eastern and western shores of the bay. Loyalist raiders operated mainly out of Tangier and Hog Islands but often took shelter in the lee of Watts Island.

When war came again in 1812, British forces used Watts as a base from which

*Right:* Chart of Watts Island, 1905. (Courtesy of National Archives (NACP) RG 23, T2695)

*Far right:* A 1652 grant surveyed Watts Island at 400 acres. It had eroded to 92 acres in 1998, the slow but steady loss averaging 1.5 acres per year. In 2002 a three-foot shoal south of the island marked the former site of the Watts Island light. (NOAA 12228, 2002. Courtesy of Maptech, Inc.)

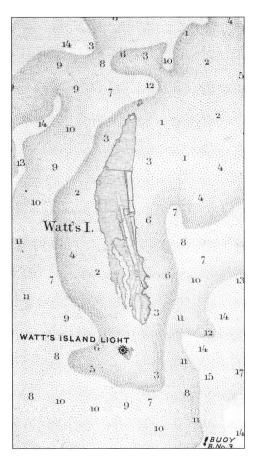

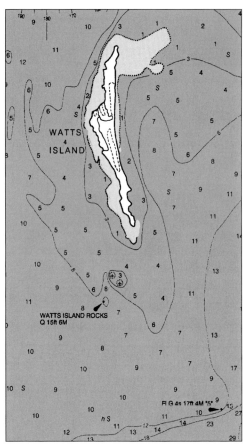

to send out raiding parties and control the lower Chesapeake. They built a fort on the island, which was patrolled by a small naval force commanded by Adm. Robert Barrie in the HMS *Dragon*. A story is told of the American brig *Eliza,* whose captain attempted to slip through the Tangier Straits undetected by the British ships off Watts Island. Enemy lookouts spied him, and several vessels gave chase. The captain only had time for a quick look, but he thought he saw a frigate lying in the channel off Watts Island. He didn't know that the *Dragon* was not a frigate or that he'd nearly come under the guns of a powerful and heavily armed line-of-battle ship.

Relative peace descended on the bay, where the federal government had plans for a seven-acre islet off the southern tip of Watts Island. Known as Little Watts, it was government property, and in 1833

entrepreneur John Donahoo accepted a commission to build a forty-eight-foot tower and twenty-by-thirty-four-foot light keeper's house for $4,755.

Meanwhile, a small village of about two dozen houses, a small church, and a store had grown up on Watts Island. With the outbreak of the Civil War, passing Union and, occasionally, Confederate steam-driven boats found Watts a convenient source of firewood to keep their boilers going. By the end of hostilities, foraging parties had cut down nearly all of the island's trees.

Shortly after the Civil War, the beacon on Little Watts Island received a new, stronger lamp. The next improvement came in 1891 when the federal government's newly created lighthouse board authorized expansion of the one-story lighthouse keeper's cottage to better accommodate keepers and their families.

By 1893 the house had a second story and a yard enclosed by a white picket fence.

Originally, Watts and Little Watts Islands and the rich oyster beds around them were claimed by Maryland, but the actual state boundary running through Pocomoke and Tangier Sounds had never been clearly defined. That lack spawned bitter disputes between Maryland and Virginia oystermen during the Oyster Wars of the 1870s and 1880s. When the states established a boundary in 1877, it did little to improve relations between their watermen. The decision gave most of Pocomoke and Tangier Sounds to Virginia, barring Marylanders from fishing in those waters. Watts and Little Watts officially became part of Virginia but changed little else in the life of the islanders.

Of greater import was the island's continuing erosion. One by one, family by family, Watts Island residents began to leave. Almost all were gone by 1908 when the island's owner, Arintha Doremus, sold it to Dr. Daniel Hardenberg for $726. His brother, Charles, arrived two years later with several horses, farming tools, and cases of books and lived on the island completely alone for ten years. As a result, he came to be known as the "hermit of Watts Island." When his horses died, Charles abandoned farming for fishing and crabbing. Only once was he in trouble. One winter, ice stopped all traffic on the bay and ended the delivery of food and other necessities to Watts Island. Aware of his plight and his refusal to leave, some Tangier Islanders crossed the three miles of ice pulling a sled loaded with supplies. At

The waters of Tangier Sound threaten to engulf the 111-year-old lighthouse on Little Watts Island during a storm in the fall of 1944. The beacon collapsed a few weeks later. (From a U.S. Coast Guard photo)

the end of ten years, Charles returned to his native Jersey City. He stayed away only long enough to find a wife before returning to the bay, this time to Little Watts Island as light keeper.

Little Watts was down to three acres by then, and high tides and storms brought the waters almost to the Hardenbergs' door. When the light was automated in 1923, they moved to Watts Island. A terrific hurricane that swept the bay in 1933 forced Charles and his wife to flee to the relative safety of the mainland. After the storm, Charles returned alone to live on Watts. When, in 1943, the two islands passed out of Hardenberg hands, they were abandoned to the elements. Just a year later, a winter storm severely undermined the foundations of the light tower and keeper's house on Little Watts, and a few weeks later the lighthouse collapsed.

In 1966 Randolph Kleinfelter bought Watts Island at an auction for $6,500. At the time wildlife was still abundant, especially herons. The loblolly pines were filled with nesting eagles, great and little blue herons, egrets, Louisiana herons, and night herons. Thousands of other migrating birds stopped over. In addition to raccoons, opossums, and foxes, about 125 goats lived on the island. The goats survived until 1978 when vandals slaughtered most of the herd. The diminishing habitat gradually reduced the rest of the animal population, and salt encroachment into the water table began to kill the trees, reducing the nesting sites for herons and eagles. In 1995 the U.S.

Fish and Wildlife Service purchased Watts Island as a wildlife refuge. The better part of the 91.71-acre island—80 percent—was classified as wetlands in 1998. Little Watts was only a shoal marked by a lighted buoy, a memorial of sorts to the hermit of Watts Island, the last human being to call either island home.

REFERENCES

1652 Waddelow grant (400 acres); 1670 Augustine Herrman Map, 1678 John Thornton and Robert Green Map, 1685 Christopher Browne Map, 1776 Anthony Smith Map, Papenfuse/Coale; 1850 TC, USCGS, 309 (310.56 acres); 1905 TC, USCGS, 2695 (253.92 acres, Little Watts 6.36 acres, total 260.28 acres); 1942 Tangier Topographic Map; 1968 Tangier Topographic Map; 1984 NOAA 12228, (Watts 104.58 acres); 1998 NOAA 12228 (91.71 acres). Erosion loss 1.48 acres per year.

Burgess, Robert, *This Was Chesapeake Bay* (Cambridge, Md.: Tidewater, 1963; reprint, 1983). Cordry, Mary, "Island-Loving Pair: Couple Owns Two in the Lower Chesapeake Bay," *Sun*, January 11, 1981. Eller, Ernest McNeill, ed., *Chesapeake Bay in the American Revolution* (Centreville, Md.: Tidewater, 1981). Hayes, Anne H., and Harriet R. Hazleton. *Chesapeake Kaleidoscope* (Cambridge, Md.: Tidewater, 1975). Hornberger/Turbeyville. Kleinfelter, G. R., interview by the author, January 13, 1987. Marriner, Kirk, "The Hermit of Watts Island," *CBMag* (February 1983). Scarupa, Henry, "Eastern Shore by Coach & Four," *Sun Mag*, September 14, 1980. Shomette, Donald G., *Pirates on the Chesapeake* (Centreville, Md.: Tidewater, 1985). Whitelaw, Ralph T., *Virginia's Eastern Shore: A History of Northampton and Accomack Counties,* vol. 2 (Richmond: Virginia Historical Society, 1951).

Deal Island, not to be confused with the town of Deale on the Western Shore, lies in Tangier Sound in Somerset County. About 54 percent of Deal Island is marshland, and most of the residents live along Deal Island Road, which follows the high ground that runs more or less down the center of the island. Deal, one of the two towns on the island, is near the northern end of Deal Island Road. Wenona is toward the south. The life of these towns is centered around crabbing, oystering, and other water-oriented enterprises. The islanders are almost all working watermen, following a centuries-old tradition. No one lives on marshy Little Deal Island.

Artifacts found on Deal Island suggest it was the site of a Manokin Indian village. As so often happened, they left soon after the English settlers arrived, leaving only shards of pottery, arrowheads, and bits of stone tools as a sign of their early presence. A little more than thirty years after the founding of Maryland, Lord Baltimore issued land grants on the island. One of the earliest settlers was John Laws, an entrepreneur who encouraged others to settle there and work the surrounding waters. He envisioned a good living to be had from the sale of fish, oysters, and crabs.

In the eighteenth century the island was a lawless refuge for pirates and earned the name Devil's Island. Even after the pirates had been routed from their lair, the island remained a hideout for those intent on mischief. During the American Revolution, Tories used it to stage raids on shipping in Tangier Sound and the Lower Bay and continued their raiding after most hostilities had ended. In March 1783 Maryland put together a

*Top:* Chart of Deal Island (here marked as "Deil's Island"), 1849. (Courtesy of National Archives (NACP) RG 23, T270)

*Bottom:* Deal Island's acreage dropped from 2,280 in 1948 to 1,950 in 1998, averaging a loss of 6.6 acres per year. Little Deal Island eroded from 368 acres in 1948 to 197 acres in 1998, a relatively greater loss of 3.4 acres (well over 10 percent) a year. (NOAA 12230, 2002. Courtesy of Maptech, Inc.)

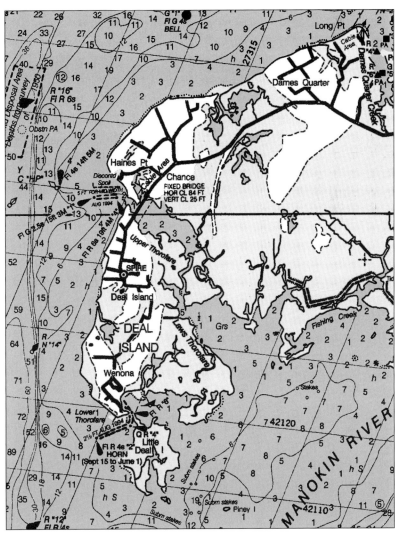

*Above:* Oyster shells piled on Deal Island testified to the productivity of its packing house in the 1950s. (Bodine Collection, The Maryland Historical Society, Baltimore, Maryland)

*Opposite:* In 1953 Aubrey Bodine photographed Melvin Collier, a local blacksmith, and a few of the oyster dredges he maintained. Bodine Collection, The Maryland Historical Society, Baltimore, Maryland)

force of soldiers and sailors aboard the barges *Fearnaught* and *Defense* and the schooner *Venus* to attack the Tory raiders at their base on Devil's Island and "seize all hostile ships in that area." The small force caught the loyalists by surprise and "captured a large quantity of the enemy's plunder." What became of the captured plunder, no one knows. The expedition against Devil's Island was the last action of the Maryland Navy, which was soon disbanded.

Religion came to Deal Islanders in the form of Methodism. According to legend, an elder of the congregation changed the island's name from Devil's, or Deil's, to Deal. The first of several famous preachers was David Wallace, a Methodist minister who arrived in 1744. His home, known as "Support," served

as the first church until it was destroyed by fire. Later, Capt. William Price bought the property and built a new house, which still stands and is a private residence.

The most famous of those who brought the word of God to islanders around Tangier Sound was Joshua Thomas, the "Parson of the Islands." He began holding camp meetings on his native Tangier Island in the 1820s and then moved to Little Deal Island in 1825. He led the first camp meeting on Deal in 1828. More than two hundred sailboats and a steamboat or two brought worshipers from miles around to gather in the wooden shelters set up for the occasion. The grounds were lit at night by fire stands, which were bonfires set atop tall piles of mud and sand. Enterprising

salesmen set out their wares for the islanders to look over—if they could tear themselves away from the preaching, which apparently they did.

Thomas carried the gospel and other preachers to communities on lower Eastern Shore islands and the mainland in his log canoe, the *Methodist*. He was an elder of the Methodist congregation on Deal Island when the islanders built a frame chapel in 1850. When he died three years later, Joshua Thomas was buried in its graveyard. The little chapel still stands behind St. John's United Methodist Church, one of three Methodist churches on the island. The tradition of the yearly camp meeting continued through the twentieth century, and today they are held under a roofed pavilion named for Joshua Thomas.

From the beginning, Deal Islanders prospered from their harvests of oysters, fish, and crabs and from running packing houses, boatyards, and sailmaking operations—mainly those industries that supported the work of the watermen. By 1850 the population had grown to five hundred. Before the Civil War, Wenona sailmaker John Stubbs opened a sail loft that became one of the island's principal businesses. Stubbs eventually sold the loft to Henry Brown, whose descendants ran it for many years as Albert Brown & Brothers. They were famous for their hand-finished custom sails for pleasure boats, and watermen swore by Brown sails for their skipjacks. In the 1980s the last Brown retired, abandoning the aging frame loft building in Wenona. Almost gone are the skipjacks. Recently a small fleet of about a dozen moored between the mainland and Deal Island in the narrow strait known as Upper Thorofare—as opposed to Lower Thorofare, which separates Deal and Little Deal Islands. They were among approximately 150 skipjacks that worked

the bay in the 1940s. About sixty were actively dredging in Maryland in 1960 when a group of skipjack captains from Deal Island established an annual Labor Day race. By the 1970s the number of skipjacks had dropped to about fifty. In the late 1980s the number was down to eight, and in 1998 no more than two or three dropped anchor in the Lower Thorofare. The entire Chesapeake Bay skipjack fleet numbered twelve, the last of 1,500 skipjacks, bugeyes, sloops, and schooners dredging oysters one hundred years earlier. Many islanders now operate charter boats and take out hunting parties in the fall and winter, but in 2002 Deal Islanders owned five skipjacks, the *Fannie L. Daugherty, Ida May, Somerset, City of Crisfield,* and *Caleb W. Jones.*

In 1870 five general stores served the residents of Deal Island. Work-boat builders were kept busy at two boatyards, one at the northern end of the island and one on the lower east side. An itinerant boat builder and several individuals also built skipjacks, including the *E. C. Collier,* now part of the educational program at the Chesapeake Bay Maritime Museum in St. Michaels, and the *Minnie V.,* a floating classroom operated by the Maryland Historical Society.

*Above:* St. John's United Methodist Church, Deal Island, 1988. (Author photograph)

*Opposite:* Oyster boats at Deal Island, circa 1961. The tableau included patent tongers in foreground, with skipjacks and a buy boat in rear. (Bodine Collection, The Maryland Historical Society, Baltimore, Maryland)

Memories are all that remain of oyster-shucking houses on Little Deal and tomato-canning operations on Deal Island. The soft-shell-crab industry, however, has provided a steady livelihood, with several firms still operating today.

In the early years, a ferry carried people from the mainland to Deal Island across the Upper Thorofare. A rough plank bridge replaced the ferry in the late 1880s, and in the early 1900s a more substantial wooden bridge was built across the Upper Thorofare. The growth that had brought the change from ferry to bridge also ushered in a new age of prosperity. In 1881 the Maryland Steamboat Company built a one-quarter-mile-long wharf to accommodate the steamboats that were part of the Baltimore, Chesapeake, and Atlantic Railway system, established in 1894 as the steamboat company's successor. The complex that accompanied the wharf included a warehouse, a ticket agent's office and waiting room, and, at one time, an oyster-packing house. The Anderson Hotel provided convenient accommodations for summer visitors brought in by steamboat at the turn of the twentieth century. The wharf was a great success until a hurricane in 1933 so weakened it that it had to be abandoned a couple of years later.

In addition to extensive bulkheading, the harbors of both villages were dredged in 1987. Today the channel through the Upper Thorofare runs four feet deep to Chance and, due to silting in the Lower Thorofare, only two and one-half feet deep to Wenona. The two islands have not escaped the effects of erosion. In 1948 Deal Island was 2,280 acres and Little Deal was 376 acres. By 1998 Deal was down to 1,951.85 acres, a loss of 6.56 acres a year, and Little Deal was just 197 acres, a loss of less than 2 acres per year.

REFERENCES

1849 TC, USCGS, 268; 1901 TC, USCGS, 2575; 1903 Deal Island Q, USGS; 1942 Deal Island Q, USGS; 1948 MdGS (2,280.1 acres); Little Deal Island (376.7 acres); 1972 Deal Island Q, USGS (2,098.08 acres); 1973 Terrapin Sand Point Q, USGS; Little Deal Island (270.97 acres); 1984 NOAA 12231 (2,045.74 acres); Little Deal Island (211.72 acres); 1998 NOAA 12231 (1,951.85 acres); Little Deal Island (197.38 acres). Erosion loss Deal Island 6.56 acres per year; Little Deal Island less than 2 acres per year.

Blair, Carvel Hall, and W. D. Ansel, *Chesapeake Bay Notes and Sketches* (Centreville, Md.: Tidewater, 1970). Brugger, Robert J., *Maryland, A Middle Temperament, 1634–1980* (Baltimore: Johns Hopkins University Press, 1988). Eller, Ernest McNeill, ed., *Chesapeake Bay in the American Revolution*, Maryland Bicentennial Bookshelf (Centreville, Md.: Tidewater, 1981). Evans, Ben, *Memories of Steamboats, Camp Meetings, Skipjacks, and the Islands of the Chesapeake* (private printing, 1977). Footner, Hulbert, *Rivers of the Eastern Shore* (Centreville, Md.: Tidewater, 1964). Long, Myra Thomas, *The Deal Island Story* (private printing, n.d.). "Low Tides Idle Skipjacks in Somerset Fishing Village," *Sun Mag,* January 2, 1986. *Maryland's Historic Somerset* (Princess Anne, Md.: Somerset County Board of Education, 1969). *New Guide.* "New Post Office on Deal Island Gets Stamp of Approval," *Sun,* August 22, 1991. Stiverson, Gregory, *An Island in the Chesapeake: Documenting Deal Island History* (Annapolis: Maryland Hall of Records, 1984). Truitt, Charles J., *Breadbasket of the Revolution* (Salisbury, Md.: Historical Books, Inc., 1975). Wennersten, John R., *The Oyster Wars of Chesapeake Bay* (Centreville, Md.: Tidewater, 1981).

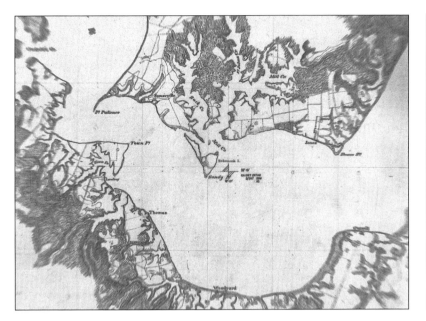

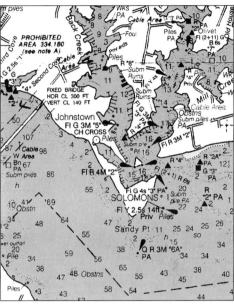

Solomons Island fell within the domain of the Patuxent Indians, who lived by farming, hunting, and fishing and had more then twenty villages along the Patuxent River. They do not seem to have settled on the island but most likely camped there to hunt or fish. By the 1650s most had died of diseases brought by the English settlers. The first Englishmen to claim land in this section of Calvert County were Edward Eltonhead and his nephew William. The two were encouraged to immigrate to the Maryland colony in 1649 by Cecil Calvert, second Lord Baltimore. Their initial grant was a manor of 2,000 acres. William was one of the supporters of the government at St. Mary's City who fought in the Battle of the Severn in 1655. He was captured by the Puritan forces of Providence in Anne Arundel County and executed. His uncle apparently survived and in 1662 was promised a grant of 5,000 acres on the condition that he bring fifty settlers into the colony. He failed to meet this condition,

even with a time extension. Ownership of the land reverted to Lord Baltimore, but not before Edward had established its identity as Eltonhead Manor.

In 1740 the island was known as Somervells Island and was owned by Dr. James Somervell, a Scottish prisoner taken by the English at the Battle of Sheriff Muir in 1715 and transported to the colonies a year later aboard the ship *Good Speed.* He had two sons, Alexander, who became a sheriff and justice of Calvert County and served in the French and Indian War, and John, who also was a justice. Both represented the county in the General Assembly.

The island had two other names in the first half of the nineteenth century— Bournes Island for Samuel Bourne, a Patuxent customs inspector, and the more widely known Sandy Island. Perhaps the most significant event in the life of the island was its purchase by Isaac Solomon in 1865. He married a young woman from the Somervell family, and they moved into the old Somervell

*Left:* Chart of Solomons Island, 1848. (Courtesy of National Archives (NACP) RG 23, T256)

*Right:* Commercial and residential investment at Solomons Island has led to its gaining 2.6 acres over the years. In 2002 it occupied 52 acres. (NOAA 12264, 2002. Courtesy of Maptech, Inc.)

house, one of two on the island. Solomon built a large-scale commercial fishery and later added an oyster cannery. His operation attracted watermen from the Eastern Shore, who came to fish, crab, and dredge for oysters or to work in the cannery. They brought their families to settle on the island and also established the mainland community of Solomons. By 1870 a 550-foot-long foot bridge connected island and mainland. It was replaced with a bridge that could handle vehicles in 1907. At that time the island's population numbered 237 with fifty-one households.

When John S. Farren and Thomas Moore bought the oyster house in 1879, fishing and oystering were prosperous businesses. In 1880 the oyster fleet working out of Solomons exceeded 500 vessels.

In 1885 when M. M. Davis took over the shipyard that Isaac Solomon had opened in 1859, he established Solomons Island as a center for boat building. The Davis yard built schooners, the workhorses of the bay; pungies, the final incarnation of the Baltimore clipper; and bugeyes for the oyster fleet. James T. Marsh of Solomons Island is credited with building the first framed and planked bugeye, the *Carrie*. Yards at Solomons Island produced at least fifty-five bugeyes. When the demand for work boats diminished, the yard began building pleasure boats, producing a number of famous racing yachts and the popular Cruisalongs.

Another Marsh, blacksmith Charles L., recorded a first when he invented the so-called patent tongs in 1887. With these tongs, oyster tongers, who once were limited to shallow waters, could

*Above:* Clifford Lusby, an oyster tonger, plies his trade in the mid-twentieth century. (Courtesy of Calvert Marine Museum, Solomons, Maryland)

*Opposite, top:* The Narrows, a scene of bovine activity before construction of the causeway in 1912–1915. (Courtesy of Calvert Marine Museum, Solomons, Maryland)

*Opposite, bottom:* Visitors to Solomons—and there were many—could stay at the Locust Inn and frolic for the photographer. (Courtesy of Calvert Marine Museum, Solomons, Maryland)

Waterside view of the J. C. Lore & Sons Oyster House, 1936. Built in 1934, this packing plant became a National Historic Landmark and, as a property of the Calvert Marine Museum, is open to the public. (Bodine Collection, The Maryland Historical Society, Baltimore, Maryland)

reach the deeper beds like those near the mouth of the Patuxent.

At one time a small, man-made island occupied the middle of the harbor at Solomons. Created in 1973 on the site of a vanished natural island known variously as Molly's Leg, Mol's Leg, and Maleg Island, it was built up with spoil from the dredging of the harbor. Before its demise, the harbor island had a colorful history. At one point watermen used its ten-foot cliff face for drying their fishing nets. A local family grazed their horses on the island, and until the

Solomons Marine Hospital closed, it used the island as a burial ground for sailors who died with no known kin. In more recent times, the islet was a haven for nude sunbathing.

In 1925 Dr. Reginald Truitt of the University of Maryland selected Solomons Island as the site for the first permanent state-supported marine biological laboratory on the East Coast. The operation expanded in 1973 to include a modern controlled marine-development laboratory, with office and laboratory spaces and increased research facilities. Known today as the Center for Environmental Science (CES), it has a fleet of research vessels equipped with a wide variety of sampling instruments. The center studies are involved with environmental problems, fisheries and shellfish biology, and chemical effects—particularly pollution.

Established in 1975, the Calvert Marine Museum is on the nine-acre site of the Solomons schoolhouse. It emphasizes local marine history, the paleontology of the twelve-to-fifteen-million-year-old fossil-laden Miocene Calvert Cliffs, and the estuarine biology of the Patuxent River and the Chesapeake Bay. Moved to the museum grounds from the mouth of the Patuxent River in 1975, the restored Drum Point Lighthouse is one of five surviving low screwpile lighthouses in the United States. Visitors may also visit the 1930s-vintage J. C. Lore & Sons Oyster Packing House and ride on the 1899 *Wm. B. Tennison,* the oldest certified passenger-carrying vessel on the bay.

Early in the twentieth century, the navy began to play a part in the life and economy of the Solomons Island area. During World War II nearby waters were used for mine testing, and the shores of Cove and Drum Point prepared thousands of marines for their landings on Pacific island beaches. Directly across the river is the Naval Air Warfare Center, and farther south is the Patuxent Naval Air Station. A great change in the Solomons area economy occurred in 1977 with completion of the towering 140-foot-high Governor Thomas Johnson Bridge across the Patuxent River. Before the bridge, which is named for Maryland's first governor, people living on or around the island and working at the Patuxent Naval Base had a long drive or crossed the river on a ferry that was formerly a fishing boat. Seafood statistics indicate the increase in the island's economy that occurred with the advent of the navy. In 2000 Solomons Island watermen caught 12,504 pounds of fish, 154 bushels of oysters, and a combined total of 3,721 soft- and hard-shell crabs.

Recreational boating and a variety of water-related festivals and events sponsored by the Calvert Marine Museum and local organizations contribute significantly to the economy and life today on Solomons Island. The island attracts visitors who enjoy walking or biking through a community rich in history, old houses, restaurants, and especially marinas. Every summer an estimated 2,600 boats tie up in the marinas and at moorings in adjacent creeks.

## REFERENCES

1848 TC, USCGS, 256; 1905 Solomons Island Q, USGS; 1907–08 TC, USCGS, 2681; 1941 HC, USCGS, 6684; 1942 TC, USCGS, 8543; 1944 Solomons Island Q, USGS (49.58 acres); 1998 NOAA 12264 (48.96 acres). Erosion loss 0.11 acre per year.

Arnett et al. Beitzell, Edwin W., *Life on the Potomac River* (Abell, Md.: private printing, 1968). Brooks, Kenneth F., Jr., "An Island of the Mind," *CBMag* (September 1974). *Calvert Marine Museum* (Solomons, Md.: Calvert Marine Museum, n.d.). *Chesapeake Biological Laboratory* (Solomons Island, Md.: Chesapeake Biological Laboratory, n.d.). Eshelman, Ralph G., and Clare Dixon, *Historical Tours through Southern Maryland: Solomons by Foot, Bicycle, or Boat* (Solomons, Md.: Calvert Marine Museum, n.d.). Manchester, Andi, "Solomons," *CBMag* (August 1985). Meyers, Karen, "The Picture Postcard Island," *Maryland Magazine* (Spring 1985). Papenfuse et al. Sjoerdsma, Ann G., "In Solomons, Today Meets the Past," *Sun,* October 14, 1985. Stein, Charles Francis, *A History of Calvert County, Maryland* (Baltimore, Md.: Schneiderith and Sons for Calvert County Historical Society, 1960).

BROOMES ISLAND
*Fossils & Fisheries: Reflections of the Past*

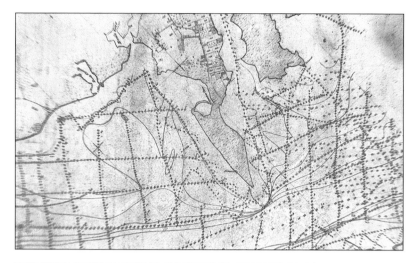

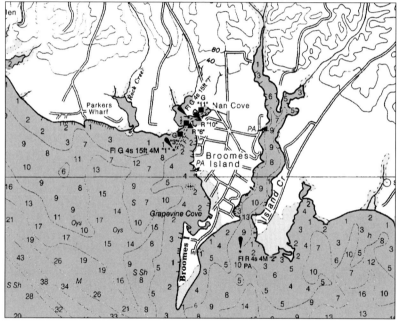

*Top:* Chart of Broomes Island, 1864–1904. (The date range reflects details incorporated during a series of surveys conducted over 40 years.) (Courtesy of National Archives (NACP) RG 23, T812)

*Bottom:* Because of its relatively sheltered location, Broomes Island has lost only about nine acres in the past forty-seven years. (NOAA 12264, 2002. Courtesy of Maptech, Inc.)

By water Broomes Island lies twelve miles up the Patuxent River just above Island Creek. This spit of land with its extending shoal reaches more than halfway across the river. On land, travelers follow Maryland Route 264 through great stretches of Calvert County countryside where farmers have devoted much of their acreage to tobacco. The highway runs the length of Broomes, which is not really an island but a peninsula of about sixty acres. Almost directly

across the Patuxent from Broomes is Drumcliff. The rugged bluff presents a rare look into the prehistoric past where erosion has exposed a geologic formation of Miocene fossils. Barnacle, pecten, oyster, and other bivalve fossils stud the crumbling face of the cliff, which is also a repository for Maryland's state fossil, the reddish brown, spiral-shaped *Ecphora Quadrisostata.*

Across the river, the earth of Broomes Island preserves ancient Indian artifacts rather than fossils. A thousand years before Christ, Indians must have hunted on the peninsula, leaving behind stone tools, broad-based arrowheads, and other artifacts. When Capt. John Smith made his map of the Patuxent in 1612, he noted nearby Indian villages but none on Broomes Island. He identified one village site across Island Creek as Opament and another on Nan Cove at the upper end of the peninsula as Quomacac. The main Patuxent Indian village was a few miles upstream.

English occupation of the island dates to 1651, when Lord Baltimore granted a 2,750-acre tract including the island to John Brome. An innholder and merchant, Brome came to Maryland at the suggestion of Thomas Cornwallis, one of the first colonists. Bromes Manor was one of many English manors along the Patuxent. The Bromes were an old family, reportedly descended from Fulk Brome, Count of Anjou, the last Crusader King of Jerusalem. Accounts say that he wore a sprig of brome flower in his helmet. John Brome's son, John Jr., was born in Maryland in 1676. He became a planter, married, and had three sons and five daughters. The younger Brome represented Calvert County in the lower house of the Maryland Gen-

eral Assembly and later served as county sheriff. He died in 1739.

The island remained in this family for several generations. John Brome III held several civil and military posts and, sometime before his death in 1748, changed his name to Broome. His son, John Broome IV, commanded the Calvert County Militia in the French and Indian War. Fighting in Western Maryland in 1755, he won great distinction as a frontier warrior. During the American Revolution, John Broome V served on the Committee of Safety for Calvert County. Broome organized a

local militia company and fed and equipped them at his own expense. In reprisal, legend says, local loyalists destroyed the Broome plantation near the island. A later descendant, Nathaniel, was an early developer, selling lots for new homes. After the Civil War, he established the fishing settlement now known as Broomes Island. About 140 families make up the community today.

At the heart of the community are the post office and Denton's, an old-fashioned general store. Another landmark is a property that once belonged to fishing boat captain Gourley Elliott. For

Broomes Island fishing guide Gourley Elliott took parties out in the bay for many years. This pier and shack belonged to his once-thriving business. (Courtesy of Calvert Marine Museum, Solomons, Maryland)

119

Warren Denton, owner of the
Warren Denton & Co. oyster-
packing house, hoists a pair
of excellent rockfish, about
1950. (Courtesy of Calvert Marine
Museum, Solomons, Maryland)

years the captain took out fishing parties. He eventually married Sadie Mister,
and the enterprising pair added a lunch-
room. Sadie's was a popular local eatery
with the added attraction of slot machines. Later, Sadie's became Stoney's
Restaurant, which is famous for its excellent crab cakes.

Up into the late 1940s and early
1950s, seine fishing was a lucrative
Broomes Island business. Working their
1,800-foot-long seine nets in nearby waters, the fishermen encircled entire
schools of hardheads, or croakers as they
often called them, entrapping several
tons of fish at a time. They then loaded
the fish into live boxes and returned the
nets to the water. When satisfied with
the day's haul, the seine fishermen carried their catch to Sewell's dock to un-
load them for processing. C. R. Sewell's
Marine Supply Store is all that remains
of the old loading dock.

Fin fishing was not the only local
seafood industry. Early in the twentieth
century, crab packing employed many
area residents. Warren and John Denton
began packing crabs on Broomes Island
in 1927. Until 1979 clam shucking also
contributed to the island's economy. In
that year, the Lowery Seafood Company, Calvert County's only clam-shuck-
ing and processing plant, shut down.

By the 1970s diminishing supplies of
clams, oysters, and fin fish all but elimi-
nated local seafood industries. Those employed in the few surviving fisheries take
special pride in their work. Ruth Mackall
Smith, working for the Warren Denton
oyster-packing company, became famous
for her speed at shucking oysters—as
many as twenty-two gallons a day. Recently, she and her brother, Cornelius
Mackall, focused public attention on
Broomes Island when they won the U.S.
National Oyster Shucking Championship.

Area children attended school on the

island until 1977 when it closed. The Broomes Island Wesleyan Church has ministered to islanders' spiritual needs since 1970. Before they established the present church, Broomes Islanders called their congregation the Pilgrim Holiness Church, a house of worship organized originally as Knapp's Chapel.

People on the island like to tell the story of a politician invited to speak there by his friend the preacher. The appointed summer day was bright and sunny, prompting members of the congregation to move the event outdoors into the churchyard. They brought chairs from the Sunday school and arranged them in the shade of trees near the water's edge. The preacher cautioned his friend that the people liked short speeches, but the politician ignored the warning and delivered his usual long-winded speech.

"How'd I do?" he asked the preacher when he was done.

"Terrible!" was the reply.

"How can you say that?" said the politician. "They kept coming up from the back row so they could hear better!"

"You dang fool," said the preacher, "the tide was coming in!"

The island's post office, one of the smallest in the United States, 1961. (Courtesy of Calvert Marine Museum, Solomons, Maryland)

Broomes Island watermen prop up a dead shark for the camera, 1967. (Courtesy of Calvert Marine Museum, Solomons, Maryland)

Tides have not done the damage they often do to bay-area islands, nor have storms contributed to appreciable erosion. Those blowing in from the northeast or southwest have little opportunity to build up strength over the water before reaching the island's shores. Those from the northwest do minimal damage. As a result, Broomes Island has lost less than an acre in the past 141 years. Enough remains to reward visitors with opportunities to experience a farming and fishing community that still reflects life in southern Maryland fifty and even one hundred years ago.

REFERENCES

1857 HC, USCGS, 641 (66 91 acres); 1892 Leonardtown Q, USCGS; 1901 Leonardtown Q, USGS; 1943 Broomes Island Q, USGS; 1944 HC, USCGS, 6963; 1953 NOAA 12264; 1963 Broomes Island Q, USGS; 1972 NOAA 12263; 1998 NOAA 12263 (57.99 acres). Erosion loss 0.63 acre per year.

Deats, Ken, "Old Times, Good Times, Hard Times: Ninety Years on the Patuxent," *Calvert County Life* 1 (1980). Hutchins, Ailene W., *John P. Brome: Memoirs* (Calvert County, Md.: published by the author, 1977). Johnson, Paula, *Historical Tours through Southern Maryland: Broomes Island* (Solomons Island, Md.: Calvert Marine Museum, 1983). Stein, Charles Francis, *A History of Calvert County*, rev. Bicentennial ed. (Calvert County, Md.: published by the author, 1977). Stephonaitis, L. C., "A Survey of Artifact Collections from the Patuxent River Drainage," Maryland Historical Trust Monograph Series, no. 1 (Annapolis: MdHT and the MdDNR Coastal Resources Division, 1980).

Once known as Old Island, Janes Island is in Somerset County northwest of Crisfield and is part of Janes Island State Park. The island, which is separated from the mainland by the Little Annemessex River and Daugherty Canal, can only be reached by boat. Once a ferry carried people to the island, but no longer.

The Annemessex Indians were the island's first known inhabitants, and we can only assume that the island's name comes from an early owner or resident. Although much of Janes Island is marshy, early English settlers found enough high ground to farm with considerable success. The fertile soil produced a bounty of watermelons, cantaloupes, apples, and peaches whose quality was legendary.

Nearby waters were also bountiful, supplying oysters for a growing market. The Civil War slowed but did not cut off the demand for Chesapeake Bay oysters. In 1862 a local waterman, Capt. Alex Corbin, bought a sloop and outfitted it for dredging. As the story goes, he was headed home from Baltimore with his new boat, when he decided to try out the dredges in Tangier Sound, ignoring the fact that, according to local law, it was not oyster season. No sooner had he hauled in his catch than the police steamer *Chesapeake* pulled up alongside. The officer aboard arrested Corbin but left him aboard his boat as the steamer towed it up the Little Annemessex River. Once they reached the river landing, later known as Crisfield, the dredge boat captain could expect to be fined and possibly do time in jail. He wasn't about to let either fate befall him. Biding his time, Corbin let the steamer

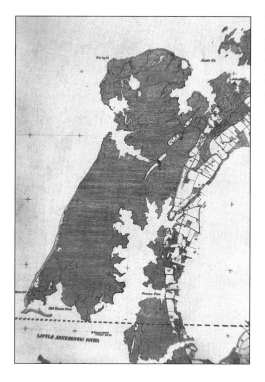

Chart dating to 1849, depicting Janes Island before excavation of the Daugherty Canal to the Manokin River in 1939. (Courtesy of National Archives (NACP) RG23, T272)

tow him up the Annemessex River until they were opposite Old House Cove on Janes Island. There, Corbin's mate cut the towline with an axe and hoisted the jib, which was all the sloop needed to sail into the shallow water of the cove. Unable to follow, the steamer stood off the entrance to the cove and fired the shots that were the opening salvo of the famous "Battle of Old House Cove." Not the least intimidated, Corbin retaliated with an old musket he had aboard, and when he ran out of musket balls, he substituted fishing sinkers. The battle continued until evening fell, at which time, Corbin and his crew swam to shore and made their way to the homes of friends. The next day, the captain left for the South, joined the Confederate Army, and fought through the war.

At the mouth of the Little Annemessex River was a treacherous sandbar upon which unwary sailors risked a

In 1849 Janes Island consisted of 3,781.4 acres; by 1998 it had lost 2,509 acres—two-thirds of its area. In a typical year over this century and a half, 17 acres washed away. (NOAA 12231, 2002. Courtesy of Maptech, Inc.)

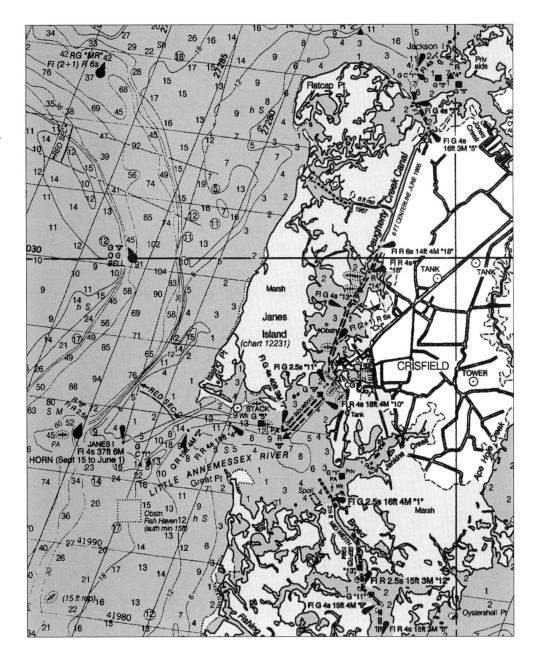

grounding. To prevent such an event, in 1853 the federal lighthouse board anchored a lightship off Island Point at the southern end of Janes Island. A second ship replaced the first in 1866 but lasted no more than a year. When the hostilities ended, the demand for Chesapeake Bay oysters regained momentum and led to the opening of a rail line to Crisfield in 1867. That in turn meant an increasing number of oyster-laden vessels were entering the Little Annemessex River

bound for Crisfield packing houses. To keep them off the sandbar, the government installed a screwpile lighthouse off Janes Island. It stood until the winter of 1899, when ice demolished it. A second lighthouse suffered the same fate thirty-six years later. Finally, in 1936 the present cylindrical caisson-type light and bell were installed to mark the entrance to the river and Crisfield Harbor.

The flourishing oyster trade after the war turned the fishing village, which by then was called Crisfield, into a rowdy boom town. That may have been why, in 1882, local authorities had to build a so-called pest house on Janes Island to quarantine and care for people with smallpox and other contagious diseases.

In 1871 on the south end of Janes Island, L. E. P. Dennis and R. H. Milligan established a factory to process menhaden for fish oil and fertilizer. Small boats trapped many thousands of pounds of menhaden in large purse nets and then transferred their daily catch to a larger boat for transport to the factory. There the fish were boiled in open kettles to extract their oils and then were dried for fertilizer. The factory burned in 1900 but was soon rebuilt. In 1908 menhaden fishing in Maryland waters was outlawed because the purse nets were catching too many edible fish. As a result, the factory closed and moved to Dymers Creek, Virginia, where it operated for two years before being relocated to Sandy Point near Cape Charles, Virginia. The factory continued to operate until World War I, when it and all the steamers were taken over by the federal government for the war effort. The company immediately moved back to Crisfield, remodeled and refitted the old factory on Janes Island, and operated until 1929. The building stood empty until 1932, when fire destroyed everything but the smokestack, which still stands and continues to be a handy landmark for boats entering the port of Crisfield.

A severe hurricane in 1933 and several subsequent storms, along with encroaching high tides, have destroyed the fine beaches and farm land on the island, where no one has lived for many years. The smokestack is the only structure still standing, and the graves of those poor souls who died in the pest house have long since been swept into Tangier Sound.

Now owned by the state, Janes Island State Park and Janes Island offer opportunities for camping, boating, swimming, crabbing, and fishing. Large areas of quiet and solitude are preserved year-round in the miles of isolated shorelines and marsh populated only by waterfowl, birds, and sea creatures.

REFERENCES

1849 TC, USCGS, 272 (3,781.39 acres); 1878 HC, USCGS 1447 a & b; 1948 HC, USCGS 7722; 1948 HC, USCGS 7778; 1968 Terrapin Sand Point Q, USGS (1,436.22 acres); 1973 Terrapin Sand Point Q, USGS; 1984 NOAA Chart 12231; 1998 NOAA Chart 12231 (1,272.19 acres). Erosion loss 16.8 acres per year.

"Janes Island Combines All of Somerset County's Natural Beauty," *Salisbury Daily Times*, June 17, 1982. *Janes Island State Park* (Annapolis: MdDNR, Maryland Park Service, n.d.). Wennersten, J. R., *The Oyster Wars of Chesapeake Bay* (Centreville, Md.: Tidewater, 1981). Wilson, Woodrow T., *History of Crisfield and Surrounding Areas* (Baltimore: Gateway Press, 1977).

# THE LOWER BAY

## THE POTOMAC RIVER AND TANGIER SOUND

# V

irginia and Maryland began on islands of the Lower Bay, one in the James River, the other in the Potomac. The first Virginians called their James River settlement Jamestown, and many lost their lives to disease, accident, and hostile Indians in the struggle to maintain this first permanent English foothold in North America. Future generations would owe a great debt to Capt. John Smith for saving the colony and leaving behind a map and notes from his exploration of the bay. Having learned from the Virginians' experience, Marylanders fared better when they landed on the Potomac island that they named St. Clement's. Their relations with the Indians were, for the most part, good. Troublemakers from England and Virginia were far more of a problem to the new settlement than native people. Religion played a significant part in the politics of both colonies and their earliest relations with each other. Maryland's Catholic Calverts and their supporters, bolstered by a strong Jesuit presence in the colony, were pitted against the Protestant and, more specifically, Puritan Virginians backed by powerful forces in England. From their island beginnings, the two colonies grew strong and increasingly

independent, a development that resulted in the difficult birth of a new nation.

On the eve of the American Revolution, Virginia and Maryland were headed by royal governors. Maryland's governor Robert Eden enjoyed a relatively amicable departure when he was forced to leave his post in June 1776. The experience of Virginia's last royal governor, John Murray, Fourth Earl of Dunmore, was a different matter entirely. In April 1775 Lord Dunmore seized all of the colony's gunpowder at Williamsburg, and under the protection of the British warship *Fowey* he issued a threat to his rebellious citizens that if they did not come to their senses he would "not hesitate at reducing their houses to ashes." His plans also included raising a "force from among Indians, Negroes, & other persons, as would soon reduce . . . this Colony to Obedience." Dunmore assembled a flotilla of about eighty vessels, mostly armed merchant ships, schooners, and sloops, backed up by a few British warships. Manned by loyalists, African Americans, and British soldiers and sailors, "Dunmore's Navy" began raiding along the Lower Bay and put Norfolk to the torch.

Lord Dunmore's early successes turned to disaster, and by the winter of 1775 his so-called army, with many loyalist families who had sought his protec-

*Previous pages:* Keeping a day's catch alive in a wooden crab float, Smith Island, 1945. (Bodine Collection, Mariners' Museum, Newport News, Va.)

tion, embarked for Gwynn's Island. By the following spring, people were dying of smallpox and other diseases, and when patriot forces began bombarding the island on July 9, 1776, Dunmore loaded his remaining supporters aboard the vessels of his ragtag flotilla and sailed to St. George's Island. The loyalist force built a breastwork there and spent much of that month raiding the countryside for food and water. At one point about one hundred Maryland militiamen waded across to the island where they destroyed water casks and filled up a well. Cannons from the *Fowey* finally drove them off. Conditions only worsened, and in August Dunmore and his followers set sail for England.

Islands of the Lower Bay, particularly St. George's, continued to serve as shelters and bases for loyalist activity, most particularly raids against shipping on the Potomac and the bay. To put these Tories out of action, in May 1781 the Maryland General Assembly ordered the evacuation of all islands below Hooper Strait. Islanders had to leave their boats, which the state would sell, but could take whatever else they owned. Nothing was to be left that would aid the Tory raiders. The effort to strip the islands was not entirely successful, and a year later loyalists attacked the brig *Ranger*, trading out of Alexandria. Several men were killed and buried on St. George's Island.

Although the American Revolution had ended at Yorktown in October 1781, British loyalists from Lower Bay island communities and mainland towns claimed several victories. One of their bases reportedly was on or near Tangier Island. In May 1782 Commodore Thomas Grason of the Maryland State Navy set out with three sailing barges to "seize all the inhabitants of . . . Tangier

Islands . . . with all their vessels, Boats and canoes, and other Property." On May 10, the tiny fleet came up against five Tory barges and several other vessels in Tangier Sound. Grason was killed along with several of his men, and one of his vessels was lost. The Tories continued their raids into the following year.

During the War of 1812, British troops raided several islands, dealing a blow to the economy of St. George's Island from which it never fully recovered. They also seized and fortified Tangier Island before sailing up the Potomac River to attack Washington. The Civil War brought blockade runners seeking the shelter of Lower Bay islands as they carried supplies and new recruits to the army of the Confederacy. War over oyster grounds turned Pocomoke and Tangier Sounds, rivers that emptied into the Lower Bay, and the Chesapeake itself into a battlefield where watermen fought each other and all fought Maryland's oyster police. The Oyster Wars were a deadly period in the bay's history and in the history of many islands whose residents depended on oysters for their living. Events from that time created many a legendary figure and added a wealth of colorful tales to hand down from generation to generation of islanders and other bay watermen.

From Lord Dunmore's last stand on Gwynn's and St. George's Islands to Tippety Wichety's nineteenth-century bawdy house to Cobb Island, where the spoken word was successfully sent and received by a radio-telephone for the first time, islands of the Lower Bay and the Potomac River are rich in history. Unlike Smith and Tangier Islands, where the aura of bygone days still lingers, several islands have become comfortably modern residential communities and popular summer getaways.

## SMITH ISLAND
*Sheltering Wildlife and Local Tradition for Centuries*

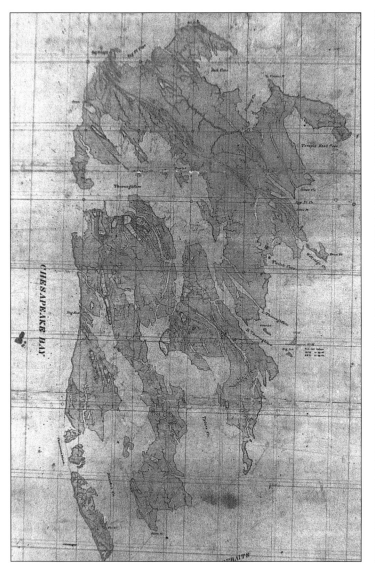

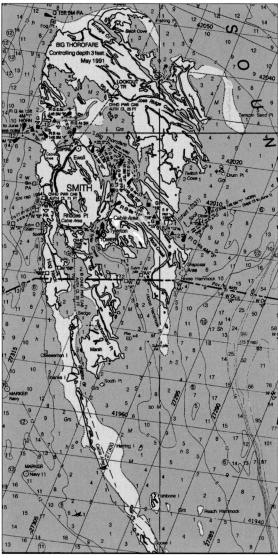

*Left:* Chart of Smith Island, 1849. (Courtesy of National Archives (NACP) RG23, T271)

*Right:* Smith Island has lost 277 acres in the past 150 years, dropping from 9,776 to 9,499 acres, a loss typically of two acres per year. While this rate is comparatively low, a great deal of land once used for raising crops now exists only as marshland. (NOAA 12230, 2002. Courtesy of Maptech, Inc.)

Eight miles west of Crisfield, Smith Island is really a collection of large and small islands defined by dozens of thoroughfares, channels, guts, ditches, and coves. Some of the islands are only a few feet above sea level, creating a watery landscape nearly all salt marsh and meadow. The entire group covers an area about eight miles long and four miles wide.

Capt. John Smith discovered these islands and those now known as Tangier during his 1608 exploration of the bay and named them "Russels Isles" for Walter Russell, the physician who accompanied him. Smith Island's actual founding is said to go back to a group of 1657 settlers who crossed the bay from St. Mary's City. The present name is not in honor of John Smith but of Capt. Henry Smith, a colonial planter on the mainland who owned a large part of the island. In those early years, mainland farmers like Smith used the island for grazing their livestock and growing vegetables.

According to Chesapeake Bay author

Tom Horton, whose *Island Out of Time* takes readers deep into day-to-day life on Smith, the island's population grew slowly. From about 25 families in the late 1700s, it increased to 300 individuals at the time of the Civil War and about 800 in the early 1900s. By the end of the twentieth century, few more than 400 people made their homes on Smith Island. Marshall, Bradshaw, Tyler, Corbin, Marsh, Laird, Smith, Tull, Clayton, and Evans were common names among the island's earliest residents, and descendants of many of these first families still live on the island.

The seventeenth-century English character of island life was still evident in the customs and patterns of speech well into the twentieth century. Electricity, telephone, and most other conveniences have come to Smith Island, but in important ways, islanders live much as they did one or even two hundred years ago. They still know they can rely on their mutual trust of each other without benefit of local government, a police force, or a jail.

Their faith in God is deep-rooted. Joshua Thomas, who as a young man was known by fellow islanders for his dancing and singing, settled down and became a Methodist preacher. Remembered today as the "Parson of the Islands," he was largely responsible for the strength of Methodism that has prevailed to the present day among island residents. They still hold yearly camp meetings at the old camp grounds.

Most of the islanders live in three villages, Tylerton, Ewell, and Rhodes Point. In 1670 Tylerton's founding father, John Marshall, settled in an area called Drum Point. His descendants still live on Smith Island, as do those of the first Tylers. The latter family gave their name to the creek leading to Drum Point, and many years later, when the

village got a post office, its name was changed to Tylerton. Recent changes in population resulted in the closing of the village school, and now children take a school boat to Crisfield. Ten miles north of Tylerton, Ewell was on a tract once called Fog's Point. According to local lore, the third settlement was called Rogues Point, so named for eighteenth-century pirates known to frequent the area. Today the community is known as Rhodes Point.

The Glenn L. Martin Wildlife Refuge

*Top:* The vista the town of Ewell, Smith Island, presented from the water, 1948. (Bodine Collection, Mariners' Museum, Newport News, Va.)

*Bottom:* A residential street in Ewell, 1948. (Bodine Collection, Mariners' Museum, Newport News, Va.)

*Above:* Skipjacks at anchor, with the Ewell United Methodist Church in background, 1950. (Author photograph)

*Opposite:* Packing blowfish *(above)* and barrels of the fish *(below)* awaiting shipment. (Bodine Collection, The Maryland Historical Society, Baltimore, Maryland)

occupies the northernmost section of Smith Island. Its 4,423 acres provide a safe feeding and resting place for migrating ducks and geese and a nesting habitat for a wide variety of other waterfowl, including wading birds, shorebirds, and ospreys. The waterfowl population peaks between mid-December and January, with more than 3,000 ducks and 1,000 geese using the refuge. Herons and egrets are numerous, creating a large and active rookery in the island sanctuary. Narrow, shallow channels limit access to the refuge, which is closed to the public without permission.

Crabbing, fishing, and oystering are still Smith Islanders' main source of income, although not as much so as in the nineteenth and early twentieth centuries. In 1986, area watermen harvested four million pounds of hard crabs valued at $1.5 million from Tangier Sound and surrounding waters. They brought in almost 500,000 pounds of soft crabs with a value of $900,000, a large share of which went to Smith Islanders. In the 1990s, 95 percent of the entire nation's supply of soft crabs was caught within a fifty-mile radius of Smith Island. Totals for 2000 were 13,228 pounds of fish,

1,365 bushels of oysters, and 193,353 pounds of soft- and hard-shell crabs.

Although residents are used to a steady flow of visitors, they have resisted any all-out promotion of tourism. The Chesapeake Bay Foundation maintains a study center at Tylerton where individuals or organizations may have an overnight adventure exploring the island's ecosystems. The community has recently opened the Smith Island Center, a museum whose exhibits tell of the island's history and life. One or two small inns and a few bed-and-breakfasts cater to overnight visitors, but day-trippers are still in the majority. For those who want to visit this unique island community, several cruise boats run between the island and nearby Crisfield or Reedville in Virginia.

Everyone who has been to Smith Island has his or her favorite story, and so do I. As a staff oceanographer with Johns Hopkins University's Chesapeake Bay Institute, I was aboard the new research vessel *Maury* when she tied up at the public dock for her first visit to Smith Island. A number of islanders watched us with evident interest, but not a soul spoke. Finally, one old man could contain his curiosity no longer.

"What kind of a boat is that?" he asked.

We invited him aboard to find out, and that was all it took to break through the other islanders' natural reserve. Within ten minutes about twenty-five people had come aboard armed with questions.

My brother, the late Dr. Eugene Cronin, used to tell another story involving the installation of a weather station on the island and the announcement that someone was needed to run it. The first candidate was enthusiastic about the opportunity until he discovered that the position was strictly volun-

teer. He quickly lost interest. In doing the job, that is. When a volunteer finally came forward to run the station, the man who'd refused the job regained his interest. He was quite willing to volunteer his advice.

"Why don't they put up that flag, don't they know it's going to blow?" he might say, or "Here it is on the ebb tide and slick cam [calm], why don't they take down that flag?"

As another story goes, an overabundance of cats was, and still is, a perpetual problem on the island, and a group of residents decided to attempt a solution. They enlisted the aid of the *Island Belle I,* the only boat running to the island at the time, and rounded up all the cats they could catch. They loaded them into cardboard cartons and onto the *Island Belle* for shipment to Crisfield, a place with plenty of fish and crab scraps to support a few more cats. The boxes were securely tied to the top of the boat's cabin and would have made it across Tangier Sound without incident if the *Island Belle* hadn't run into a sudden squall. In no time the driving rain had turned the cardboard boxes to mush, and the cats were all over the boat. Inevitably, some went over the side, and small heads bobbed behind the boat like crab pot buoys. Most, it seems, survived and swam ashore on Janes Island. The majority remained aboard for the rest of the trip, but as soon as the boat touched land, they made a beeline for shore. What the citizens of Crisfield made of the sight was never recorded.

Not counting cats, the island's population has been declining steadily. It stood at about 375 at the turn of the twenty-first century, which was down from 675 in 1980. Nature has had a hand in the changes occurring on Smith Island. Erosion is the culprit. In 1849 the island encompassed 9,766 acres of mostly fields and woods; in 1998 most of its 9,449 acres was marshland. The highest land is just three feet above low tide, and a high tide of the full moon will completely flood the lowlands. Adding to the problem, the bay's water level is rising at a rate of about one-fifth of an inch a year, and the lowering water table has resulted in significant saltwater infiltration. The combination spells doom for the island. To help control land loss, in 1994 the Army Corps of Engineers put several one-hundred-foot-long, five-foot-thick bags filled with dredge spoil around Rhodes Point, which was eroding at a rate of almost eight feet a year. Thus far, the erosion rate has been stemmed, but there is no certainty for the future.

REFERENCES

1849 TC, USCGS, 271 (9,766.19 acres, three large, eight small islands); 1856 HC, USCGS, 5571; 1879 HC, USCGS, 1441 a & b; 1901 TC, USCGS, 2556; 1917 Ewell Q, USGS; 1942 Ewell Q, USGS; 1951 HC, USCGS, 7943; 1967 Ewell Q, USGS; 1972 Kedges Straits Q, USGS; 1973 PR Ewell Q, USGS; 1998 NOAA Chart 12201 (9,499 acres). Erosion loss 1.79 acres per year.

"Airstrip, Town Houses proposed for Smith Island," *Sun*, n.d. "Glenn L. Martin Wildlife Refuge," U.S. Department of the Interior (n.d.). Greiser, Bob, "Smith Island, Tranquility and Tradition Threatened by the Bay," *CBMag* (June 1978). Grossfield, Stan, "Rising Winds, Tides Threaten Smith Island," *Sun*, February 8, 1998. Horton, Tom, *An Island Out of Time: A Memoir of Smith Island in the Chesapeake* (New York: Vintage, 1996). "Island Erosion Control Planned," *Annapolis Capital*, April 23, 1998. *Maryland's Historic Somerset: Smith Island* (Princess Anne, Md.: Board of Education of Somerset County, 1969). *Maryland Travel Guide* (Baltimore: Maryland Office of Tourist Development, yearly). Middleton, Alice Venable, *Maryland's Right Tight Island: Smith* (private printing, n.d.). Mountford, Kent, "Our Island in the Chesapeake," *CBMag* (August 1978). A Portrait of Smith Island," *Sun*, April 24, 1986. Rodgers, William H., "Smith Island: A Trip Back in Time," *Sun*, April, 20, 1986. Sayles, Tom, "The Immutable Smith Island," *Mid-Atlantic Country Magazine* (February 1989). "Smith Island: 'A Wet and Windy Kingdom,'" *Maryland Magazine* 11, no. 3 (1979). "Smith Island is a Refuge for Great Blue Herons: Could it Be One for Watermen?" *Sun*, June 1, 1986. Spangler, Todd, "Trying to Shore up Smith Island," *Mid-Atlantic Country Magazine*, May 10, 1998. Tyler, Capt. Otis Ray, interview by the author, April 1987.

## TANGIER ISLAND
*Brine-soaked Tradition and a Changing Way of Life*

Like so many of the islands in the Lower Bay, Tangier is really a group of several islands. They lie ten miles from the nearest point on the Eastern Shore and fourteen miles from Virginia's mainland to the west. Though part of Virginia, Tangier Island's closest link is with Crisfield in Maryland, one and a quarter hours away by boat. From north to south, the islands occupy a five-mile stretch of water at the entrance to the sound that bears their name.

Only one of the islands—the one generally referred to as Tangier—is inhabited. It is a mile long and only seven feet above sea level at its highest point. Seventy-nine people lived on the island in 1800: thirty-three Crocketts, twenty Evanses, and twenty-six others. By 1890 the population was up to one thousand. It peaked in the 1930s at 1,500 and has been diminishing slowly ever since. At present approximately 700 people live on Tangier.

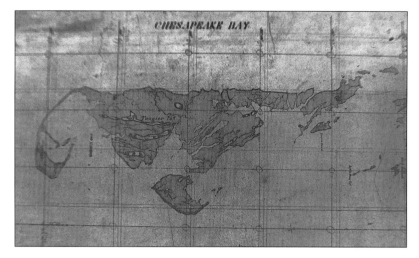

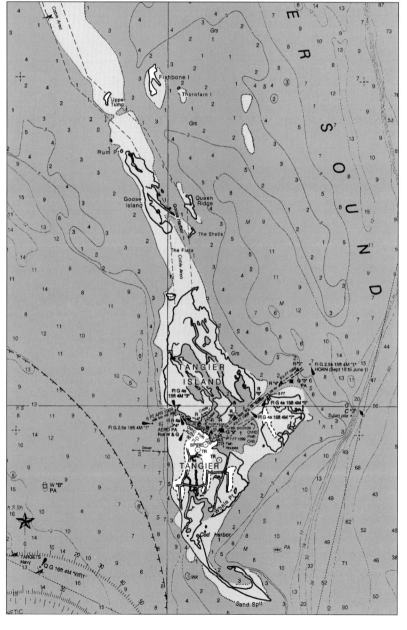

Early charts beginning in 1689 refer to it as Tanger or Tager Island but offer no clues as to the origin of the name. Not until 1832 is it called Tangier Island. Archaeologists have unearthed projectile points used by native peoples some ten or twelve thousand years ago, before the waters of the Atlantic had risen to drown the Susquehanna River and create the Chesapeake Bay. The natives undoubtedly left their arrow- and spearheads on what was then the mainland. Other archaeological finds include points and scrapers used by Indians who lived at the time that the first English settlers arrived.

When making the map of his exploration of the Chesapeake Bay in 1608, Capt. John Smith designated these islands and those of Smith Island in a group he named "Russel's Isles" for the physician accompanying his crew. The first permanent settlement was in 1686 when John Crockett arrived with his eight sons and their families to farm and raise livestock. Besides the trials nature imposed on those who would live on an island, the Crocketts had to contend with raids by crews of passing vessels who regularly came ashore to forage for fresh food, took anything they could, and butchered the family's animals.

Although the hostilities between Britain and her American colonies had ended in 1781, British loyalists continued to attack ships in the lower Chesapeake, reportedly from bases on or near Tangier Island. In the spring of 1782, Commodore Thomas Grason of the Maryland State Navy set out with three sailing barges to "seize all the inhabitants of . . . Tangier Islands . . . with all their vessels, Boats and canoes, and other Property." On May 10, the tiny fleet came up against five Tory barges and several other vessels in Tangier Sound. Grason died in the battle along with several of

REV. JOSHUA THOMAS PREACHING TO THE BRITISH ARMY ON
TANGIER ISLAND 1814

his men, and one of his vessels was lost. The Tories continued their raids into the following year, which was also the year that the Treaty of Paris marked the official end of the American Revolution.

Islanders enjoyed relative peace until the War of 1812, when about 1,200 British troops descended on Tangier Island. They set up a three-sided fort with large twenty-four-pound cannons for defense of the island, which they used as a base for their attack on Baltimore. While there, the British soldiers listened to the preaching of Joshua Thomas, the Parson of the Islands, and most likely scoffed when he predicted that they would not take Baltimore. The attack on Fort McHenry, during which Francis Scott Key wrote our national anthem, did fail, and the British left the island in defeat. The fort and graveyard they left behind are now under ten feet of water one-half mile out in the bay.

The years that followed were gener-

ally more tranquil but not without crises. In 1821 a hurricane hit with destructive force. Disease also took its toll. In 1866 the entire population of the island was evacuated to Crisfield during a deadly cholera epidemic. Both tuberculosis and measles struck in the 1870s, and in the 1880s islanders fought off smallpox.

Tangier Island lay at the entrance to one of the world's most prolific oyster grounds, and in 1867 when the railroad reached Crisfield, islanders shared in the resulting economic boom. In addition to oysters, the market for fish, crabs, and clams flourished. The watermen of Tangier Island were ideally situated to help supply the demand, which remained strong into the early 1920s. As the supplies of fin and shellfish dwindled, the island boom towns of Oyster Creek, Canaan, and Rubentown struggled to survive. However, they were done in by nature rather than the econ-

*Top:* Tangier Harbor in the mid-twentieth century, a scene of perfect serenity. (Chesapeake Bay Maritime Museum, photograph by Constance Stuart Larrabee)

*Bottom:* Harvesting eelgrass, with a beached and abandoned menhaden fishing boat in the background. (Bodine Collection, The Maryland Historical Society, Baltimore, Maryland)

*Opposite:* A waterman poses for Aubrey Bodine's camera with bags of eel grass, to be used as packing material for soft crabs. (Bodine Collection, The Maryland Historical Society, Baltimore, Maryland)

Enjoying a meal in a work
boat galley, about 1950.
(Bodine Collection, The Maryland
Historical Society, Baltimore,
Maryland)

omy. One by one, they succumbed to
the rising waters of the bay. For the
same reason, islanders who raised livestock on the once rich island meadow-
land began to cut back their herds.
Today none remain.

In 1911 the U.S. Navy towed the bat-
tleship USS *San Marcos*, formerly the
USS *Texas*, to a spot about six miles

southwest of the island for target practice. The USS *New Hampshire* opened
fire, hitting her forty-two times before
she sank. Awash at low tide, the ship's
hulk caused at least seven shipwrecks
over the ensuing years. Finally, in the
1950s dynamite was used to dig a trench,
and the *San Marcos* rolled over into it
and today lies at a depth of twenty feet.

The walk to Swain
Methodist Church, ca.
1950s. (Chesapeake Bay
Maritime Museum, photograph
by Constance Stuart Larrabee)

Another wreck caused considerable excitement in the 1920s. A strong northwest wind swept the water away from the island's western shore, exposing the wreck of an old sailing vessel. Islanders found engraved copper dishes and a brass axe blade of Spanish origin. The wreck added a new tale of treasure to other local legends, one of which involved Blackbeard, the notorious pirate, who is said to have buried a hoard of his ill-gotten gains on the nearby island of Shank's Hammock.

Nature struck Tangier Island with a vengeance in the 1930s. The great hurricane of 1933 created waves that sent spray two stories high on island houses and carried away much of the shoreline. Three

years later, the bay froze, and the islanders had to depend on Red Cross food and supplies dropped by planes. Tangier Islanders still depend on help from the air for medical care. In the 1990s Dr. David B. Nichols and a nurse flew in from Whitestone on the Western Shore. Appointed by the U.S. public health service, such doctors offer free health care to those watermen who qualify.

The airstrip is one of the busier places on the island. As many as seventy-five planes land every week bringing hundreds of tourists. Visitors are charmed by the strong sense of the past that still exists on Tangier Island. Long-time residents have adapted slowly to modern ways, retaining a unique inner grace,

Tangier's oldest cemetery, 1988. (Author photograph)

along with customs and the distinct island brogue of their ancestors. Picket fences may have given way to chain link, and electricity and telephones are everywhere, but island streets are still narrow, and most islanders still make their livings working the water.

Catering to the needs of vacationers, as well as a number of environmental scientists, has also provided a living for Tangier Islanders. Covering 250 acres, Port Isobel Island to the east of the main island was once a vacation spot for swimming, crabbing, fishing, and hunting. Today, thanks to the generosity of G. Randy Kleinfelter, the island belongs to the Chesapeake Bay Foundation and is used by students, teachers, and others to study its extensive natural and cultural resources.

The natural resources of Tangier and its sister islands are in grave danger. Muskrats still make their homes in the marshes, rails and herons stalk the shallows, and in the fall wild ducks and geese from the north still descend on the islands to winter, but erosion threatens the land and the habitats it supports. Between 1850 and 1997, the loss of high ground has averaged nearly 8.98 acres a year, creating large tidal flats that are exposed at low tide and turning much of the remaining land into marsh. The next major hurricane could be a very serious threat not only to flora and fauna but to houses and businesses as well.

Despite the long-term problems they face, Tangier Islanders continue to live as they have for generations, adapting as conditions dictate, and accepting the steady flow of visitors between April and November. In addition to planes that bring them to the island, several cruise boats make once-a-day trips from Crisfield in Maryland and Reedville and Onancock in Virginia. Many visitors spend the night to better experience life on this unique Chesapeake Bay island.

REFERENCES

1659 John Thornton & William Fisher Map, 1692 Jacobus Robyn Map, 1696 Pierre Mortier Map, 1708 Herman Moll Map, 1714 J. B. Homan Map, 1721 Herman Moll Map, 1737 Homan Heirs Map, 1751 Joshua Frye & Peter Jefferson Map, 1776 Anthony Smith Map, 1747 Emmanuel Bowen Map, 1781 Robert Alexander Map, 1794 Dennis Griffith Map, 1794 Captain N. Holland Map, 1832 Fielding Lucas Jr. Map, Papenfuse/Coale; 1849 HC, USCGS, 252; 1850 TC, USCGS, 309 (2,060.02 acres); 1856 HC, USCGS, 557; 1905 TC, USCGS, 2695 (1,627.27 acres); 1911 HC, USCGS, 3361; 1917 Ewell Topographic Map; 1937 TC, USCGS, 5681; 1942 TC, USCGS, 8164; 1951 HC, USCGS, 7944 (above Oyster Creek Gut, 583.99 acres, Tangier Island proper 506.72 acres, East Island 141.25 acres, Goose Island 115.91 acres, total 1,412.48 acres); 1956 HC, USCGS, 8407; 1968 Tangier Island Q, USGS (nine islands 1,054.35 acres); 1984 NOAA 12228 (Upper Island 2.55 acres, Fishbone Island 7.65 acres, Goose Island 76.52 acres, Queen Ridge 2.55 acres, Upper Tangier 372.42 acres, Lower Tangier 441.28 acres, East

Tangier 142.84 acres, total 1,045.81 acres); 1997 NOAA 12228 (solid ground 82.95 acres, marshland 656.29 acres, tidal flats 743.65 acres, total ground 739.24 acres). Erosion loss 8.93 acres per year.

Bradshaw, Vernon, interview by the author, May 9, 1988. Duke, Maurice, "The San Marcos Wreck," *CBMag* (November 1983). *CBAm Rev.* Frye, John, "Tangier's Flying Doctor," *CBMag* (n.d.). Harr, Dorothy, *Eastern Shore by Coach and Four* (Cambridge, Md.: Tidewater, 1976). Novak, Josephine, "Escape for a Day, Not Far Away, to Isle in Bay," *Sun*, June 26, 1981. Painter, Deborah, "The Mammoth Hunters of Tangier," *CBMag* (October 1987). Russell, Linda L., "Tangier: Island in the Bay is an Island in Time," *Sun*, September 20, 1967. Steuart, Lenmann, and Vernon Bradshaw, *Visitors Guide to Tangier Island* (private printing, 1976). Timberg, Craig, "An Island Split Over Money, Sin," *Sun*, March 14, 1998. Trevillian, R., and F. Carter, *Treasure on the Chesapeake Bay* (Glen Burnie, Md.: Spyglass Enterprises, 1982). "Two-Hundred-Fifty-Acre Island Donated to Environmental Group," *Sun*, 1988. Wheatley, Harold G., "This Is My Island: Tangier," *National Geographic* (November 1973). Wilson, Woodrow T., *History of Crisfield and Surrounding Areas* (Baltimore: Gateway Press, 1977).

## TIPPETY WICHETY ISLAND
### *A Small Jewel in the St. Mary's River*

Tippety Wichety Island has had a colorful history and several name changes. Because land records before 1827 were lost when the St. Mary's County Courthouse burned, earlier information on the island is difficult, if not impossible, to find. Records may not exist, but ancient pottery shards and arrowheads found on the island are evidence of Indian occupation. Early names included St. Marie's Island and Lynch Island for its owner, Stephen Lynch, a St. Mary's County tobacco inspector.

After the Civil War, Lynch Island became something of a legend, not the least part of which was the origin of its new name. As one story goes, an alleged ex-Confederate smuggler, Capt. H. W. Howgate, was running one of a number of floating bordellos anchored off the city of Alexandria. He had a successful business, to the dismay of Alexandrians, until one spring in the late 1870s, the Potomac flooded and swamped or washed away most of the bordellos.

Alexandria's city fathers had been waiting for just such an opportunity to rid their waterfront of the disreputable establishments and ordered the remain-

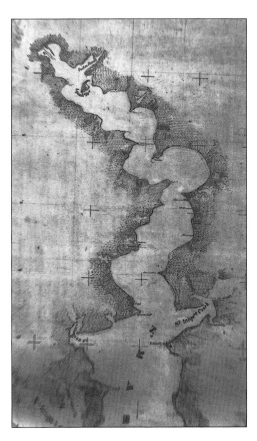

Chart of Tippety Wichety Island (then Lynch's Island), St. Marys County, 1858–1904. (Courtesy of National Archives (NACP) RG 23, T776, NARA)

ing owners to leave. Among them was Capt. Howgate, who owned several floats. Undaunted, he began looking for another site for his activities and discovered that Lynch Island was for sale. He

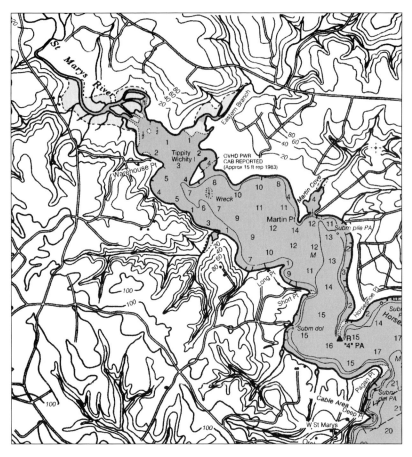

In 1858, Tippety Wichety Island claimed 7.64 acres; in 2002 its size had dropped to 5.1 acres. Erosion has been less than .02 acres per year, one of the lowest rates in the bay. (NOAA 12233, 2002. Courtesy of Maptech, Inc.)

bought it in 1879 for $300, built a house with three upstairs bedrooms, and opened for business. His girly, gambling, and drinking place was known as "Happy Land," or in some circles, the "Tippling and Witching House." Out of that eventually came the name Tippety Wichety. However enticing, Howgate's new location was apparently too remote to make a profit. He finally sold the island in 1881 for $5,600. In addition to the island's name, Happyland Road, about two miles south of Great Mills, is the only surviving evidence of this particular chapter in the island's colorful past. Howgate's house survived until World War II, when it was destroyed by vandals.

Other sources of the name Tippety Wichety have been suggested. One that's been circulating for some time claims that one night, under a full moon per-

haps, a man in a canoe passed by the island on his way home from a drinking spree. Although he was a bit tipsy, he was managing quite well until he looked across the water to the island. There he saw an apparition, looking for all the world like a witch, and she was pointing directly at him. Suddenly, he lost control of the canoe, which tipped him into the river. Of course, no one believed him when he got back to shore, but his story was a good one, and soon people were calling the island "Tip o' the Witch." Perhaps the same witch had something to do with the one-hundred-foot sailing barge that sank southeast of the island and is a navigational hazard today.

The existence of the bawdy house, at least, was confirmed by Mrs. Margaret Lewis of Fairfields, a farm adjacent to the island. Apparently, an acquaintance told her that his aunt was the madam at the Tippety Wichety house. She personally recalls visiting the island as a child for picnics and rowing over to gather flowers from around the deserted house. She and her friends also visited the island in the winter to gather bittersweet to decorate their homes for Christmas.

In 1974 Earnest Dickey bought the island and built a home on the site of the Tippety Wichety house. To Margaret Lewis's dismay, he had a small herd of goats that completely consumed the bittersweet bushes. Today the island belongs to John Harman and Richard Timble, who live there with their families and a large black dog who thinks that the island belongs to him. The owners have planted grasses along the island's shore to retard erosion. They are aided in their efforts by Tippety Wichety's sheltered location and weak tides that have kept erosion much lower than is usual for bay islands. The land loss has averaged only 0.02 acres per year.

REFERENCES

1858 TC, USCGS, 776; Lynch Island (7.64 acres); 1942 St. Mary's City Q, USGS (7.35 acres); 1942 TC, USCGS, 8138; 1943 St.

Mary's City Q, USGS (6.42 acres); 1960 HC, USCGS, 8548 (5.1 acres); 1984 NOAA 12233 (5.1 acres); 1999 NOAA 12233 (5.1 acres). Erosion loss 0.02 acres per year.

## ST. GEORGES ISLAND
### *From Tumultuous Past to Peaceful Present*

Lying between St. Georges Creek and the Potomac River is a well-populated island with a long and eventful history. From the arrival of the first settlers, the strategic advantages of St. Georges Island's location near the mouth of the Potomac and St. Mary's Rivers has attracted military and commercial interest, which on numerous occasions has led to outright conflict.

St. Georges Island was settled first by Ferdinando Polton, a Jesuit priest who arrived in Maryland in 1639. The island was part of the 3,000-acre St. Inigoes Manor granted to the Jesuits by Cecil Calvert, second Lord Baltimore. At that time, the island encompassed more than 700 acres. Father Polton traded with the Indians, farmed, and raised cattle on the island, which remained the property of the Jesuits until 1862.

War came to St. Georges Island in July 1776. That month Virginia's last royal governor, Lord Dunmore, arrived with what remained of the loyalist force he'd gathered to subdue rebellious Virginians. His forces built breastworks there and raided local farms for food and water. Maryland militiamen attempted to sabotage the loyalists' efforts but were driven off by the British warship *Fowey*. By August, Dunmore realized his cause was hopeless, and he and his followers left for England. A large number of loyalists remained and continued to harass "rebel" shipping in nearby waters well after British general Cornwallis's surrender to George Washington at Yorktown.

During the War of 1812, British troops again landed on St. Georges Island. They rounded up the livestock, cut timber, and then burned every house, causing fires that spread from one end of the island to the other. Pleas for military assistance were dispatched to President James Madison in Washington. His answer was that he could not afford to protect "every Southern Maryland turnip patch." Damage to the island was so great that more than 300 families left. Many moved to Kentucky, where they found a new home in Bardstown. A few people remained on St. Georges Island and in the 1820s attempted to raise merino sheep. An infestation of ticks and mosquitoes decimated the flock, putting an end to the enterprise.

During the Civil War, blockade runners used the island, with its easy access to the bay, as a base of operations. At one point, the USS *Jacob Bell* descended on St. Georges and captured some thirty Confederate soldiers encamped there. In another wartime incident, the USS *Tulip*, a Union gunboat, passed close by on her way to the Washington navy yard for repairs. Completely disregarding his orders, the boat's captain fired up the defective starboard boiler, which blew up, almost totally destroying the *Tulip* in a tremendous explosion. Only five of the crew of fifty survived.

In 1862 Ennis Rozelle and John Robrect of Tangier bought St. Georges Island from the Jesuits for $10,000. Rozelle set up a sawmill and found a ready market among island shipyards who supplied dredge boats and other

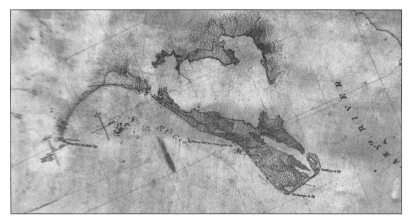

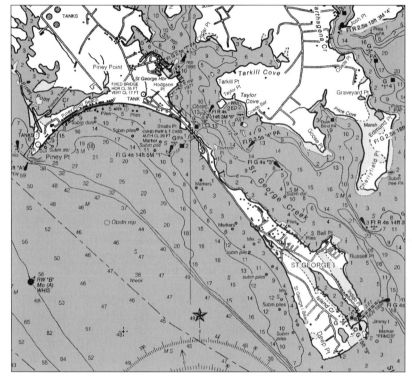

sort. Hotels sprang up along Hobbs and Adams wharves and offered vacationers a room and meals for a dollar a day. The resort business experienced a sudden growth spurt in 1891, when word spread that three crew members from the river steamer *Sue* were searching the river north of St. Georges Island for a cache of more than $40,000 in gold coins. They were "very uncommunicative" when they returned several days later, and soon St. Georges Island was swarming with treasure seekers. To this day no one knows if the men from the steamer ever found what they sought. No one else did. If any riches were to be had, they went to the hotels and other businesses that catered to the visiting throngs.

The first bridge to the island was built in 1921 but washed away in the 1933 hurricane that completely flooded the island. A second bridge suffered the same fate, carried away by Hurricane Hazel in 1954. It was replaced by the present more durable steel and concrete bridge that serves the growing number of people making the island their home. Despite erosion that has reduced the 700-acre island to 379 acres over the last ninety years, the number of houses on St. Georges Island has gone from 40 in 1904 to more than 200 at the end of the century. Island life seems to agree with those who make St. Georges their home. Many continue working the water, bringing in 12,176 pounds of fish, 1,498 bushels of oysters, and 51,141 pounds of soft- and hard-shell crabs in 2000.

## REFERENCES

1639 St. Inigoes Manor (700 acres); 1857 HC, USCGS, 640; 1859 TC, USCGS, 804; 1902 TC, USCGS, 2598 (653.79 acres); 1912 Point Lookout Q, USGS; 1942 St. Georges Island Q, USGS (551.99 acres);

*Top:* Chart of St. Georges Island, 1858–1904. (Courtesy of National Archives (NACP) RG 23, T776, NARA)

*Bottom:* When first surveyed in 1639, St. Georges consisted of about 700 acres. It now claims only about 370, having averaged a decline of one acre per year. (NOAA 12233, 2002. Courtesy of Maptech, Inc.)

craft for the growing fleet during the oyster boom of the 1870s and 1880s. The waters around the island were a rich source of fish, crabs, eels, clams, and particularly oysters. As a result, those same waters were the scene of battle during the so-called Oyster Wars fought between watermen of Maryland and Virginia, between tongers and dredgers, and between watermen of every stripe and Maryland's "oyster police."

At about the same time, Washingtonians discovered St. Georges Island. From 1870 to 1930 it flourished as a re-

1955 TC, USCGS, 110672; 1960 HC, USCGS, 8548; 1973 *PR* St. Georges Island Q, USGS; 1984 NOAA 12233; 1997 NOAA 12233 (379.06 acres). Erosion loss 3.29 acres per year.

Beitzell, Edwin W., "The Battle of St. Georges Island," *CBMag* (July 1976). Beitzell, Edwin W., *Life on the Potomac River* (Abell, Md.: private printing, 1968). Eller, Ernest McNeill, ed., *Chesapeake Bay in the American Revolution*, Maryland Bicentennial Bookshelf (Centreville, Md.: Tidewater, 1981). Hammett, Regina Combs, *History of St. Mary's County* (private printing, 1977). Hayes, Anne, "Lord Dunmore's Floating Town," *CBMag* (November 1976). Knight, George M., Jr., *Intimate Glimpses of Old St. Mary's* (New York: Meyer and Thalheimer, 1938). Manakee, H. R., *Maryland in the Civil War* (Baltimore: MdHS, 1961). Pogue, H. R., *Old Maryland Landmarks* (private printing, 1972). Pogue, Robert E. T., *Yesterday in Old St Mary's County* (n.p.: Carlton Press, 1968). Richardson, Hester, *Sidelights on Maryland History* (Cambridge, Md.: Tidewater, 1967). Schomette, Donald, *Battle for the Patuxent* (Solomons, Md.: Calvert Maritime Museum Press, 1981). Schomette, Donald, *Shipwrecks of the Chesapeake* (Centreville, Md.: Tidewater, 1982). Tilp, Frederick, *This Was the Potomac River* (Bowie, Md.: Heritage Books, 1988). Wilstach, Paul, *Potomac Landings* (Cambridge, Md.: Tidewater, 1969).

## COBB ISLAND

*Scene of the First Successful Wireless Transmission of Speech*

Not to be confused with an island on Virginia's Eastern Shore having a similar name, Maryland's Cobb Island lies thirty-three miles up the Potomac at the mouth of the Wicomico River in Charles County.

The first owner of Cobb Island was James Neale, who immigrated to Maryland in 1635. As an early settler, he received a grant of 2,000 acres, which he called Wollaston Manor, after his boyhood home in England. Neale was a gentleman, merchant, diplomat, and planter. As an agent for Leonard Calvert, Maryland's first governor, he traveled to Boston to develop trade with that northern colony. Later, he represented Leonard's brother Cecil, second Lord Baltimore, in dealings with the Dutch in Amsterdam. Between 1644 and 1660, Neale was a factor in Spain and Portugal as well as the ambassador to those countries for King Charles II and the Duke of York. When he finally returned with his family to live in Maryland in 1660, he purchased land in Charles County. Part of the purchase was Cobb Island, so-called either because it was bought with rough-cut Spanish dollars called cobs or, more likely, because it was a lumpish bit of land, which in England would have been called a cob. Neale represented Charles County in the Maryland General Assembly between 1662 and 1666.

The island was uninhabited throughout the eighteenth century. It undoubtedly passed through several hands before 1832 and the death of Henry Hammersley, whose will recorded the bequest to his son John of "the Peninsular of Cobb Point, 342 acres, more or less." No one made a permanent home on the island until George Vickers bought the entire island in 1889, built a house, and began farming. That same year, the federal government built a lighthouse at the edge of the island to guide the heavy river traffic around the Cobb Island Bar. The lighthouse keepers surveyed a constant stream of boats of all sizes and descriptions passing on the river.

At one time steamboats regularly served businesses and residents along the Wicomico River, and hundreds of freight schooners stopped at river landings to load local tobacco, oysters, and

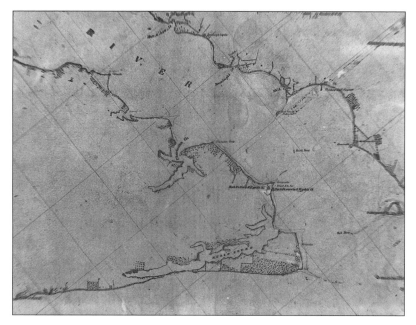

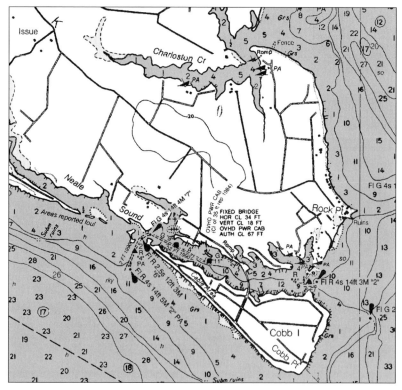

*Top:* Chart of Cobb Island, 1862. (Courtesy of National Archives (NACP) RG 23, T859, NARA)

*Bottom:* Cobb Island has lost 86 acres since 1860. About three-fourths, or 294 acres, remains. (NOAA 12286, 2002. Courtesy of Maptech, Inc.)

produce bound for markets up and down the bay. Foul weather often drove vessels into the anchorage formed by the Cobb Island Bar, where they were safe from summer storms and winter's drifting ice. The bars on both sides of the river's mouth may have offered protection, but they tested a sailor's navigating

skills. Frequently, while trying to avoid one bar, a boat grounded on the other. Sailing vessels seldom attempted entering or leaving the river at night, even after the light was installed. The lighthouse keepers witnessed countless groundings—and more.

At the height of the Oyster Wars, the oyster-rich Cobb Island Bar was often a battleground as its supply of the desirable bivalve steadily diminished. The keepers at the light reported "whenever oyster dredging is going on . . . there's sure to be floating dead bodies seen every week towards the end of the season." Shoot-outs between the Maryland "Oyster Navy" and outlaw dredgers occurred with terrible regularity in the late 1800s and early 1900s. Historian John Wennersten records one incident in 1906. That oyster season, the police had decided to keep the dredgers off the Cobb Island Bar. In a confrontation in St. Mary's City, Alex Harris, captain of a dredge boat, told Douglas Russel of the Oyster Navy's sloop *Bessie Jones* that the law wasn't going to keep him from pulling his scrape over the Cobb Island Bar. "The law and my 45–70 rifle can do just that," was the policeman's warning, but Harris ignored it. Using the cover of night, he headed for the forbidden oyster grounds. There, Russel found him, and standing on the bow of the *Jones,* he began firing into the dredge boat's rigging to bring down her sails. Once the schooner was dead in the water, capture would be easy, but Harris had other ideas. Although the waterman fired back, one shot from Russel's 45–70 killed Harris. A coroner's jury eventually exonerated Douglas Russel of the man's death.

In 1939 a fire of unknown origin completely destroyed the Cobb Island lighthouse. Three men tending the light jumped to safety in the water where they

were rescued by a navy picket boat. The federal lighthouse board replaced the lighthouse with the present bell and automated beacon, and the heap of stones that supported the original structure remains as one more navigational hazard in a river maze of bars and submerged wrecks, ruins, and other obstacles.

In the meantime an event had taken place that would profoundly affect the lives of people the world over. It began, with little public notice, in 1900. That year a Canadian electrical scientist, Reginald A. Fessenden, assisted by Frank

Lighthouse on Cobb Island, built in 1860, as it appeared in 1989. The house was dismantled and the light automated in 1940. (Author photograph)

Very, demonstrated the first radio-telephone to successfully send and receive intelligible speech. They were able to transmit their words by electromagnetic waves sent between two fifty-foot masts a mile apart. Their experiment involved, in part, winding coils of wire of many different diameters, a sometimes frustrating process about which Very commented "of these some are found to work better than others, but there are some things that we do not know." Fessenden kept at it, though. Later, with 150-foot towers, he was able to send his radio signals as far as Alexandria, Virginia. By 1906 he had moved his operations to Massachusetts and there developed circuitry that was later adopted and perfected by the radio broadcasting industry. A historical marker near the site of his first experiments commemorates the pioneering work that had its beginning on Cobb Island. The house that Fessenden and Very lived in still stands and is now a private residence.

Until 1912 Cobb Island belonged to the Vickers family, who finally gave up their failed attempts at farming. Robert Crain, a developer and politician, bought the island and organized the Cobb Island Development Company. By 1922 the company had built roads and had made a start on a new summer resort community. The next year Crain built a noisy one-way wooden bridge that was prone to summer traffic jams. It ran from Chigger City—a local nickname for the area—to the upper end of the island. The island's first church was St. Paul's Catholic Chapel, built in 1926. Later the Cobb Island Baptist Church replaced it as the only house of worship on the island. The post office opened in 1927, and the island became an incorporated township two years later. With much fanfare, a new bridge opened in 1932, providing the opportunity for a

community celebration featuring an auto parade, foot races, swimming and boat races, and a crab feast. In 1936 the township purchased its first fire truck, an army surplus vehicle that seemed a bargain at $500. It immediately needed repairs, however, after the engine blew up on the way to the island. Electricity reached Cobb Island in 1939.

Life on the island had its occasional drawbacks. The 1933 hurricane flooded the mainland end of the bridge, stranding residents and visitors. This was repeated during Hurricane Hazel in 1954. For years the Dahlgren Naval Weapons Station has regularly declared a section of the Potomac off the island a no-mans-land when the big guns are test fired. Designated the "Middle Danger Area," the navy patrols the area to control boat traffic during firing. One of the firing range boats is stationed off Cobb Island.

Today, the onetime summertime resort is a year-round community where residents enjoy the serenity of their natural surroundings. Tides, phases of the moon, high and low water, weather, and fishing, oystering, and crabbing regulate Cobb Islanders' activities and direct their conversation. In 2000 those working the water brought in 75 tons of fish, 508 bushels of crabs, and 5,890 pounds of soft- and hard-shell crabs.

Cobb Island is sheltered from most of the wind and wave action that causes the erosion that has greatly affected other islands. Most northwest winds come directly down the Potomac River, passing it by, and storms out of the northeast gain little strength to build damaging waves from the narrow stretch of water they cross before hitting the island. As a result, Cobb has averaged a loss of only two acres per year, going from 380.71 acres in 1860 to 294.66 acres in 1998.

REFERENCES

1832 Will of Henry Hammersley (342 acres, "more or less"); 1860 HC, USCGS, 969 (380.71 acres); 1862 HC, USCGS, 778; 1862 TC, USCGS, 858; 1902 TC, USCGS, 2599 (317.56 acres, 4 buildings); 1905 TC, USCGS, 2730; 1942 TC, USCGS, 8115; 1961 HC, USCGS, 8613; 1943 Rock Point Q, USGS (308.51 acres, 138 houses); 1974 Rock Point Q, USGS (292.90 acres, 249 houses); 1998 NOAA 12285 (294.66 acres). Erosion loss 0.62 acre per year.

Ackerman, Clara Dailey, *History of Cobb Island* (Cobb Island, Md.: Cobb Island Citizens Association, 1960). Chapman, Jo Ann Rector, *History of Cobb Island* (Cobb Island, Md.: Cobb Island Citizens Association, June 1980). Hornberger/Turbeyville. McMenamin, Bill, "Potomac Hideaway," *CBMag* (December 1988). Mullican, Grant, "Cobb Island Bridge Dedication," September 5, 1932. Norris, L. A., "The World's First Radio Program," *Yankee* (December 1965). Papenfuse et al. Parker, John, *La Plata Maryland Independent,* January 26, 1977. Stepp, John, "Cobb Islanders Still Trying for a Two-way Bridge," *Washington Sunday Star,* November 2, 1958. Turner, Merle, *Waldorf Maryland Independent,* October 2, 1981. Very, Frank, letter to his mother, 1900. Wennersten, John R., *The Oyster Wars of Chesapeake Bay* (Centreville, Md.: Tidewater, 1981).

# GWYNN'S ISLAND
## *Lord Dunmore's Final Stronghold*

Gwynn's is a large island just below the mouth of Virginia's Piankatank River between the York and Rappahannock Rivers. It is separated from the mainland by Milford Haven, which is crossed at its western end by a small swing bridge. At its lower end, Milford Haven empties into the Chesapeake above a spot known as Hole in the Wall.

An arrowhead found on the island attests to the presence of native peoples some ten thousand years ago. Piankatank Indians from a later period left a large number and variety of artifacts. In 1642 the entire island belonged to Virginian Hugh Gwynn, who moved there from Jamestown. He brought his wife, two sons, and a couple of servants and settled first on the mainland side of the island. He called that homestead Gwynneville and the waterway Milford Haven. Gwynn later built another home on the bay side of the island.

Gwynn's Island earned a place in the history books in the winter of 1775, when, after burning Norfolk, Virginia's last royal governor, Lord Dunmore, arrived with a flotilla transporting his loy-

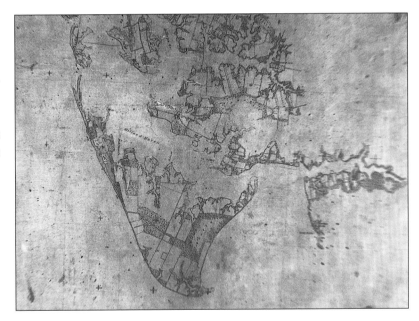

Chart of Gwynn's Island, 1853. (Courtesy of National Archives (NACP) RG 23, T503, NARA)

alist force. They set up a base of operations on the island, and soon the loyalists' families joined them. Food was scarce, and living conditions grew desperate with an outbreak of smallpox. Disease spread rapidly, and by July 1776 more than 500 had perished. Following an attack by patriotic militiamen, Dunmore's fleet set sail for St. Georges Island and by August were bound for England.

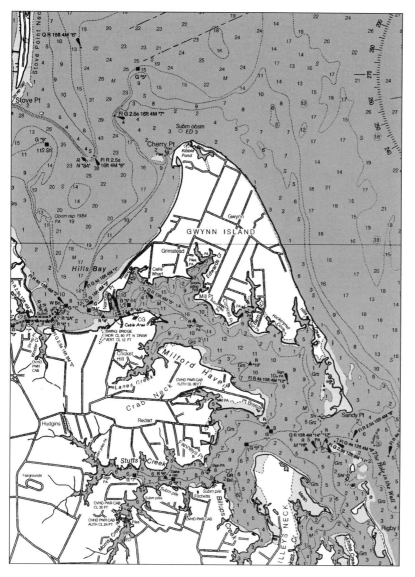

Gwynn's Island has lost 166 acres in the past 145 years, a little more, on average, than one acre per year. It is still a healthy 1,425 acres in size. (NOAA 12235, 2002. Courtesy of Maptech, Inc.)

and a stateroom. It was a sad day when the *Piankatank,* the last steamer to serve islanders, arrived in 1932 for the final trip up the bay.

Once a ferry carried people between the mainland and the island across Milford Haven, but today a highway and bridge provide faster, easier access, if not the same experience. In recent years, the population of Gwynn's Island has fluctuated, depending on the season. About 400 people live on the island year-round, but that number expands to 700 when summer comes. Fair weather brings vacationers to camp, enjoy a waterfront getaway for a day or night, and get out in their boats. In addition to fishing and crabbing, catering to visitors provides a chief livelihood for permanent residents.

## REFERENCES

1853 TC, USCGS, 503 (1,595.45 acres); 1907 TC, USCGS, 2869; 1942 Deltaville Q, USGS; 1952 TC, USCGS, 11157; 1964 Deltaville & Mathews Q, USGS (1,500.5 acres); 1998 NOAA 12235 (1,429.28 acres). Erosion loss 1.14 acres per year.

Eller, Ernest McNeill, ed., *Chesapeake Bay in the American Revolution,* Maryland Bicentennial Bookshelf (Centreville, Md.: Tidewater, 1981). Greenwood, Isaac, "Cruizing on the Chesapeake in 1781," *MdHM* 5 (1910). Hazleton, Harriet, "Marina Hopping: Windmill Point Marine Lodge and the Islander and Narrows Marina," *CBMag* (July 1977). Lanford, Charlotte, "Air, Sunshine, Legend Flavor Gwynn's Island," *Richmond Times Dispatch,* October 17, 1978. Lanford, Charlotte, "It's Gwynn's, Not Gwynn Island," *Richmond News Leader,* April 13, 1973. Sauder, Ron, "What's Not in a Name Bothers Ex-teacher," *Richmond Times Dispatch,* August 3, 1980. "Virginia Bay Village Adjusts to Newcomers," *Sun,* August 10, 1997. Walling, Eileen, "History of Gwynne's Island, 1634–1776" (research paper, George Wythe High School, 1972).

The next major event to affect life on the island occurred in 1874 when steamboat service to Baltimore began. At the time steamboats stopped regularly at bay and river landings to pick up local tobacco, oysters, and produce bound for market. Residents also made overnight trips to the city for business and to shop or attend theater performances and other events. The first steamer to pick up passengers from Gwynn's Island was the *Massachusetts,* which ran daily to Baltimore from the Piankatank and Rappahannock Rivers. The trip cost seven dollars and included four meals

*Where Marylanders First "Took Solemn Possession" of the Colony*

The history of Maryland begins on St. Clement's, but this island in the Potomac was occupied long before the English arrived. Among other evidence, archaeologists have discovered a burial cache of pre-Columbian bones. When the Maryland adventurers arrived in the *Ark* and the *Dove*, however, the island was a Piscataway hunting and fishing ground and, according to the settlers' estimate, consisted of about 400 acres.

Leonard Calvert, Maryland's first governor and brother of Cecil Calvert, second Lord Baltimore, led the group of about 140 colonists who sailed from Cowes, England, on November 22, 1633. They landed on St. Clement's on March 25, 1634, and there erected a large cross of native wood. The island was too small for a settlement, and Leonard Calvert set out with Virginia fur trader and explorer Henry Fleet to find a better setting for Maryland's first seat of government. While he was away, he ordered those left behind to build a fort, which they did.

In 1639 the island and adjacent mainland were patented by Thomas Gerard as part of St. Clement's Manor, which took in most of the land between St. Clement's Bay and the Wicomico River. Upon his death in 1673, Gerard's property passed to his son Justinian who gave St. Clement's Island to his daughter Elizabeth as part of her dowry when she married Nehemiah Blakiston in 1669. Her husband was one of the leaders in 1689 of a revolt that temporarily took control of Maryland's government out of the Calverts' hands. Blakiston became head of an interim government in 1690 and held several high offices until 1693, when charges of embezzlement of customs fees and corruption may have led

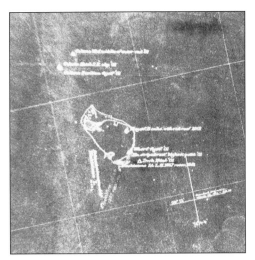

*Top:* Chart of St. Clement's Island, 1858. (Courtesy of National Archives (NACP) RG 23, T1105, NARA)

*Bottom:* Four centuries ago, an early settler estimated the size of St. Clement's at 400 acres, which may have been overly generous. A federal government survey in 1919 found just 66 acres. In 2002, the number stands at about 46 acres, so about .3 acres per year were lost in the 1900s. (NOAA 12286, 2002. Courtesy of Maptech, Inc.)

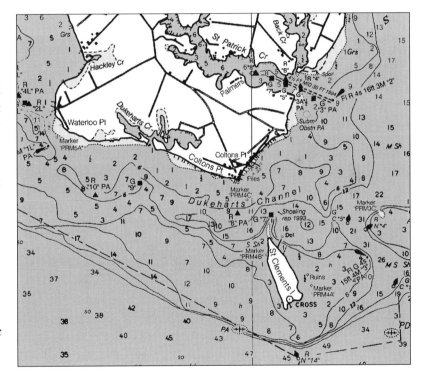

to his death that December. At various times, St. Clement's Island was also called Blakiston or Blackistone Island.

The island changed hands several times over the ensuing years until finally, in 1845, Dr. Joseph McWilliams bought it. In 1851 McWilliams allowed a two-story brick lighthouse to be built on two

St. Clement's lighthouse, 1921. The two-story house with a full front porch and bell tower was constructed in 1851. Confederate soldiers raided the house, which was repaired after the Civil War. It mysteriously burned to the ground in 1956. (Courtesy of Calvert Marine Museum, Solomons, Maryland)

acres at the southern tip of the island. Dr. McWilliams's son Jerome tended the light between 1859 and 1875. When in 1864, Confederate raiders under a Capt. John Goldsmith descended on the island, Jerome and his pregnant wife could not stop them from destroying the light and the oil that ran it, but they did persuade Goldsmith not to blow up the house. Shortly after the raiders left, Union troops arrived to secure the island and replace the light. A federal gunboat remained to patrol the area until the end of the war. Jerome's sister, Josephine Freeman, followed him and kept the light until 1911.

Joseph McWilliams developed the island after the war, adding a steamboat wharf and turning his home into a boarding house. He then added a bath house and cottages, followed by a hotel and a plan to sell lots for a town. The islanders farmed, fished, and harvested oysters from the surrounding beds, and life was good. Then, in 1889, a heavy flow of fresh water from melting snow and spring rains killed off the oysters. Prosperity soon went the way of the oysters, and the island passed through several owners who tried to make it a going

concern. In 1900, a proposed health resort to take advantage of the "Diuretic Mineral Springs of Blackiston Island" looked promising, but it too failed. Since it was first settled, erosion had also taken a terrible toll. The island that once covered 400 acres was down to 66 in 1919 when the federal government bought St. Clement's as an observation site for the navy's gun testing range at Dahlgren. The island's former owner remained to tend the lighthouse and serve as a general caretaker despite the roar of gunfire and the occasional shell hitting the island. Finally, in 1932 the light was fully automated. To commemorate the landing of the first colonists three hundred years earlier, in 1934 the state of Maryland erected a large cement cross on the deserted island. An explosion and fire, attributed to either lightning or a stray shell from the navy's nearby proving ground, totally destroyed the lighthouse and much else on the island. To clear away the dangerous ruin, the navy dynamited the remaining walls of the lighthouse.

When the navy declared the island surplus property in 1962, the state of Maryland bought it and created the St. Clement's Island State Park. Later, plans for a museum on the island as well as a replica of the lighthouse failed, but a ferry dock was built and a ferry still carries visitors to the island. In 1976 St. Mary's County established the St. Clement's Island–Potomac Museum on the mainland across Dukeharts Channel. The museum has exhibits on Maryland's first settlers, St. Mary's County history, the lower Potomac, and the Chesapeake Bay. Every July, it hosts the Blessing of the Fleet with music, tonging demonstrations, shows, and a reenactment of the landing of the *Dove* on St. Clement's Island.

## REFERENCES

1634 settlers' estimate 400 acres; 1860 HC, USCGS, 793; 1860 TC, USCGS, 1581; 1868 TC, USCGS, 1105; 1901 Piney Point Q, USGS; 1902 TC, USCGS, 2598; 1919 U.S. Government Survey (66 acres); 1943 Blackiston Island Q, USGS (61.51 acres); 1955 TC, USCGS, 10654 & TC, USCGS, 10655; 1960 HC, USCGS 8552 (55.03 acres); 1973 *PR* St. Clement's Q, USGS (48.66 acres); 1984 NOAA 12286; 1985 DNR 34-5356; 1998 NOAA 12286 (45.98 acres). Erosion loss 0.3 acres per year.

Arnett et al. Beitzell, Edwin, *Life on the Potomac River* (Abell, Md.: private printing, 1979). Brugger, Robert J., *Maryland, A Middle Temperament* (Baltimore: Johns Hopkins University Press, 1988). *Chronicles of Old St. Mary's*, St. Mary's County Historical Society, monthly. *Finer Points*, St. Clement's Island Museum (n.d.). Hornberger/Turbeyville. "Marylanders Celebrate the State's 300th Birthday," *Sun*, March 20, 1994. Papenfuse et al. Richardson, Hester, *Sidelights on Maryland History* (Cambridge, Md.: Tidewater, 1967). *Southern Maryland's Forty Museums*, Southern Maryland Museum Association (1996). Stein, Charles Francis, *A History of Calvert County*, rev. Bicentennial ed. (Calvert County, Md.: by the author, 1977). Thomas, George, *Island of Two Names*, (n.p., n.d.). Tilp, Frederick, *This Was the Potomac River* (Bowie, Md.: Heritage Books, 1988). *Walk the Paths of History and Visit St. Clement's Island*, MdDNR. Wilstach, Paul, *Potomac Landings* (Cambridge, Md.: Tidewater, 1969).

## JAMESTOWN ISLAND
*Site of the First Permanent English Settlement in North America*

This island on the James River, some thirty-one miles from Hampton Roads, is large, encompassing 978.36 acres in 1998. Today, it is a state and national park and has been preserved as one of the nation's most important colonial sites.

Nearly 400 years ago, on December 20, 1606, three small ships, the *Susan Constant*, the *Godspeed*, and the *Discovery*, set sail from Blackwall, near London, England. Aboard were 104 Englishmen and boys under the command of Capt. Christopher Newport. The voyage was financed by the Virginia Company of London, whose stockholders hoped to profit from the venture. Sailing west to "find a safe port" along the "coast of Virginia," they made landfall at Cape Henry on April 26, 1607. Passing through the Virginia Capes, they sailed straight for the mouth of a wide river they named the James in honor of their king, James I. On May 13 the adventurers finally hove to off a marshy island on the northern bank of the James and the next day went ashore. They called their outpost Jamestown or occasionally, "James Cittie," which became the first permanent English settlement in North America.

The settlers chose Jamestown because it offered good mooring and could be defended against attack by Spanish or other enemy forces approaching on the river or Indians attacking from the land. With that in mind, their first order of business was to build a fort, the site of which archaeologists recently unearthed.

Unprepared for Virginia's summer heat, native diseases, and the general hardships of establishing their precarious foothold in the New World, more than half of the first group of settlers had died by the end of their first winter. Only the action of a disciplined soldier and explorer, Capt. John Smith, kept the infant colony from total collapse. In ad-

*Top:* Chart of Jamestown Island, 1873. (Courtesy of National Archives (NACP) RG23, T1290, NARA)

*Bottom:* Surveyed in 1910 at 1,513 acres, Jamestown now runs to about 978 acres, a loss of 535 acres in 88 years, about 6 acres in a typical year. More than half of this historic island has been lost in less than a century. (NOAA 12248, 2002. Courtesy of Maptech, Inc.)

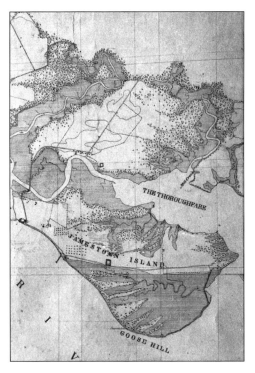

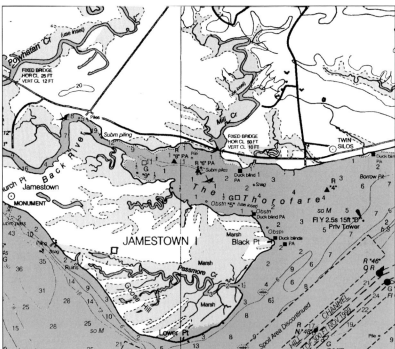

women, eight Germans, and several Poles who set up a glassworks on the island.

When Smith was badly burned in an accident and had to leave in 1609, those he left behind in Jamestown forgot or ignored the rules and practices he had set up. There was no one to keep order and maintain discipline. By Christmas most of the colonists' stored food was gone, and at the end of the winter of 1609 only 60 of the 500 had survived. In May "all save John Martin" were ready to abandon the colony when another ship arrived with supplies, more colonists, and Virginia's new governor, Lord Delaware, who refused to let them leave.

Over the following years, the colonists concentrated on their survival, building homes, planting corn, and accumulating winter stores of food. They also strove to find a source of income for the colony that would satisfy their needs and provide the profits the Virginia Company's stockholders expected. Making glass and other enterprises didn't do it. For another thirty years the colonists engaged in a vigorous and prosperous fur trade. At the same time, John Rolfe, husband of Pocahontas, began experimenting with tobacco, crossing the native plant with a less bitter West Indian variety. The result, he hoped, would find a market in England. In 1613 he sent his first harvest of the new leaf to England, and two years later, Virginians shipped 2,300 pounds of tobacco to England. By 1617 the shipment had jumped to 20,000 pounds. In time, tobacco would drive the economy and shape the life of Virginia. To raise tobacco, which required considerable acreage, the colonists began to spread out over the countryside. In 1619 twenty-eight Africans were brought into the colony from the West Indies. At first they were

dition to running the colony, Smith explored the bay, providing us with some of the earliest extant information on the Chesapeake. He became the president of the colony, which by 1609 had increased in population to about 500. Among the new arrivals were the colony's first

treated as indentured servants. Much later the system of slavery took hold in Virginia to meet the growing need for laborers in the tobacco fields.

On Jamestown Island, few settlers lived inside the fort. It was primarily a place of business and government. In 1619 Gov. George Yeardley set up the House of Burgesses, making Jamestown the provincial capital, which by the mid-1600s consisted of a brick state house and church and some thirty other buildings. Then tragedy struck. In 1622 large bands of Indians raided and killed 347 colonists, nearly a third of those living outside of Jamestown. Legend has it that Jamestown was spared because an Indian boy, who had been converted to Christianity, refused to kill his English godfather. The Englishman fled to Jamestown, warning its citizens of the attack.

Those killed in 1622 brought to 4,000 the number of settlers who had died out of the 5,000 who had come to Virginia since 1607. Back in England, administrators overseeing colonial affairs revoked the Virginia Company's charter, and Virginia became a royal colony in 1624. A concerted effort by colonial forces subsequently nearly wiped out the Indian population, ending that threat to Virginia's growth.

As for Jamestown, during its years as Virginia's capital, the town was burned four times, the last in 1698. A year later the burgesses voted to move the capital to Williamsburg, and, thanks to tobacco, the colony prospered. By 1700 its population numbered 70,000 people. Jamestown did not share in the prosperity and began a serious decline. Most of the island was used for farming, except that area around the ferry that crossed the lower James River. During the Revolutionary War, the Americans and British used the ferry. They also used the island as a place to exchange prisoners. French

troops landed there on their way to support the Americans in the war's final battle at Yorktown. Later, Confederate forces fortified Jamestown Island against Union ships using the James River.

Today, Jamestown and Yorktown make up the Colonial National Historical Park, which was created in 1930 as the first such historical park in the National Park system. At the southern end of Colonial Parkway, Jamestown Island itself is very much as the first settlers might have found it in 1607. Now a natural preserve as well as historic site, Jamestown Island offers little to suggest its early English residents. Erosion along the south shore of the island eventually swamped the early rude dwellings and the site of the first state house. The tower of the 1639 brick church, several building foundations, a graveyard, and part of the fort are all that remain of the original settlement.

Visitors must rely on their imaginations, aided by markers, artists' renderings, and audio stations, to recreate the original Jamestown settlement. Throughout the park are monuments, erected during the tercentenary celebration in 1907. Among them are statues of Capt. John Smith and Pocahontas. Frequently scheduled living history presentations help visitors understand what life was like in Jamestown. Years of archaeological digging have filled the visitor center museum with one of the largest collections of seventeenth-century artifacts in the United States. Near the mainland end of the island is the Jamestown Glasshouse, reconstructed as it might have been in 1608. Here glass blowers, using seventeenth-century methods, demonstrate their art. Nearby, Dale House is a studio where "colonial" potters work. The sale of both glassware and pottery help support operation of the park.

After seeing the actual site of Jamestown, visitors can stop at the state-run Jamestown Festival Park, which is a living history museum consisting of reproductions of James Fort, a church, an Indian village, and full-scale replicas of the *Susan Constant, Godspeed,* and *Discovery.* James Fort contains a number of wattle-and-daub buildings where reenactors show visitors what life was like in seventeenth-century Jamestown. Here weavers, spinners, armorers, basketmakers, and other artisans demonstrate colonial trades. An indoor museum offers more exhibits portraying the life of the Indians and early English settlers.

REFERENCES

1856 HC, USCGS 165; 1873 TC, USCGS 1290; 1905 TC, USCGS 2693a; 1910 HC, USCGS 3097; 1910 TC, USCGS 129Oa (1,513.35 acres); 1948 HC, USCGS 7641; 1965 Hog Island Q, USGS and Surry Q, USGS (1,135.35 acres); 1982 NOAA 12246; 1998 NOAA 122489 (78.36 acres). Erosion loss 6.08 acres per year.

Bridenbaugh, Carl, *Jamestown 1544–1699* (New York: Oxford University Press, 1980). Cotter, John, and J. P. Hudson, *New Discoveries at Jamestown* (Washington, D.C.: U. S. Government Printing Office, 1957). Fee, Robert G. C., "The Design and Construction of the Jamestown Ships" (paper presented to the Society of Naval Architects and Marine Engineers, Old Point Comfort, Va., 1958). Fishwick, Marshall, *Jamestown, First English Colony* (New York: American Heritage Publishing Company, 1965). Forman, Henry Chandlee, *Jamestown and St. Mary's* (Baltimore: Johns Hopkins Press, 1938). Hageman, James, *The Heritage of Virginia* (Norfolk: Donning Company, 1966). Haskett, James N., *Colonial National Historical Park* (Las Vegas: KC Publications, 1990). *Jamestown, Virginia, and Jamestown Festival Park* (Jamestown-Yorktown Foundation, n.d.). Tilp, Frederick, *Chesapeake: Fact, Fiction, and Fun* (Bowie, Md.: Heritage Books, n.d.). *Virginia and Maryland* (New York: Fodor's Travel Publications, 1993). *Williamsburg, Jamestown, and Yorktown* (New York: Fodor's Travel Publications, 1987).

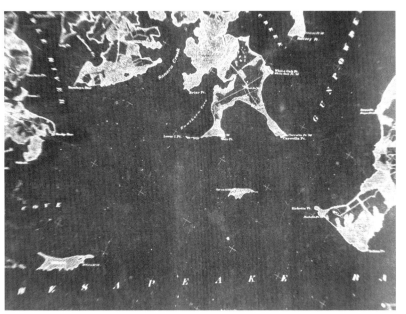

seems to have occupied the island, which was mostly marsh, and it had washed away entirely by 1974.

REFERENCES

1658 patent, part of Spry's Inheritance (640 acres); 1683 part of Maxwell's Conclusion (1,623 acres); 1807 deed, Charlotte Waltham to Samuel Rickets; 1974 (completely gone).
Marye, William B., "Early Settlers of the Site of Havre de Grace," *MdHM* 13 (1918): 198–201. Marye, William B., "Maryland Rent Rolls," *MdHM* 20 (1925): 23–24; Wright, C. Milton, *Our Harford Heritage* (Glen Burnie, Md.: French-Bray Printing Co., 1967).

*Above:* In 1846–47 a chart showed Spry's Island as a small marshy island near the mouth of Middle River. In 1845 it had an area of 89 acres, in 1933 just 9 acres, and by 1974 had totally disappeared. (Courtesy of National Archives (NACP) T213, NARA, RG23)

*Previous page:* Bodine captured this waterman and his boat at Gwynn's Island in the summer of 1943. (Bodine Collection, The Maryland Historical Society, Baltimore, Maryland)

## SPRY'S ISLAND
### *Gone from the Mouth of Middle River*

Tiny Spry's Island at the mouth of Middle River was one of those bits of water-bound land that went from owner to owner as part of larger properties. Oliver Spry patented a 640-acre tract of land known as "Sprys Inheritance" in 1658. Sprys Inheritance and the island passed from Oliver Spry to his daughter Mary Standsby in 1683 as part of "Maxwell's Conclusion," a tract of 1,623 acres. The island appears again in an 1807 deed by which Charlotte Waltham conveyed to Samuel Rickets "one third part of Maxwell's Conclusion including a small island known as Spryes Island." No one

## THREE SISTERS ISLANDS
### *Three Small Islands Long Gone*

The islands known as the Three Sisters used to lie off Horseshoe Point, just below the mouth of West River in Anne Arundel County, Maryland. They were once the hunting grounds of the Conoy Indians, a tribe allied with the Piscataways of the upper Potomac River. Edward Parrish was the first to settle a fifty-acre parcel described in 1663 as being "in the great swamp lying near the three islands." The islands were included in the patent. He owned another 100 acres, known as "Parrishes Park," on the west side of Parrish Creek. Farmers plowed and planted fields on the island in the

In the years that followed, the three islands quietly succumbed to the ebb and flow of the surrounding waters and by the 1850s had disappeared.

REFERENCES

1776 Anthony Smith Map (three islands), 1794 Dennis Griffith Map (three islands), 1840 John Henry Alexander Map (three islands), Papenfuse/Coale; 1903 HC, USCGS, 2667 (no islands).
"Correspondence of Governor Eden, 18th August 1775," *MdHM*, July 12, 1984. Fitz, Virginia, *Spirit of Shadyside: Peninsula Life* (Shadyside, Md.: PTA Committee, 1984). Shomette, Donald G., *Shipwrecks on the Chesapeake* (Centreville, Md.: Tidewater, 1982). Smith, Janet, archivist, Liverpool Record Office, personal correspondence, May 17, 1989.

Three Sisters Islands as shown in the 1878 Hopkins *Atlas of Anne Arundel County.* These three small islands were once farmed by driving horses and equipment over the shoals from the mainland. In July 1775 they supplied the scene of the indignant burning of the tea-laden merchantman *Totness*—a Maryland protest that duplicated the earlier Boston Tea Party. (Courtesy of Anne Arundel County Public Library)

early 1800s, using horses to tow their equipment across the shallows separating the islands from the mainland and each other. No one ever built houses there.

In July 1775 the 130-ton brig *Totness* went aground on one of the low-lying islands. She was apparently carrying contraband tea, which was one of the taxed commodities that American patriots had agreed not to import from Britain. Less than a year before, some of the area's more radical citizens had forced owner Anthony Stewart to burn his brig *Peggy Stewart* off Annapolis for the same offense. Local sentiment was high, and animosity grew by the hour against Liverpool merchant James Gildart who had shipped the tea, the brig's captain who had allowed it to be put aboard, and the Maryland merchants who had ordered it. The hotheads prevailed, and after allowing the brig's crew to leave with their belongings, they burned the ship to the waterline.

## SHARPS ISLAND
### *From 449 Acres to None*

Choptank Indians were probably the first inhabitants of Sharps Island west of the mouth of the Choptank River. First recorded at 449 acres, the island is now no more than a shoal. Until William Claiborne's claim was wrested from him by Lord Baltimore, it was known as Claibornes Island. Quaker doctor Peter Sharp bought it in 1675 and gave it a new name. In the following years, a number of owners farmed the island, growing wheat, corn, tobacco, potatoes, and produce.

Jacob Gibson, owner of Sharps Island in the early nineteenth century, could not keep his slaves, livestock, and crops from being taken by British raiders during the War of 1812. Nor could he escape the invaders, who took him prisoner for a brief time. Once free, he sailed jubilantly up San Domingo Creek to St. Michaels with a red banner at the peak of his vessel's mast and a boy beat-

*Right:* Chart of Sharps Island, 1848, when it occupied 449 acres and sustained four farms. (Courtesy of National Archives (NACP) RG23, T251, NARA)

*Far right:* According to this chart, Sharps in 1900 consisted of 94.2 acres, so three-quarters of the island (more than 300 acres) had disappeared in half a century. The long pier on the southeastern point connected steamboat traffic and the island hotel. (Courtesy of National Archives (NACP) RG23, T2494, NARA)

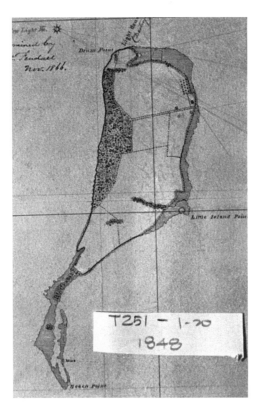

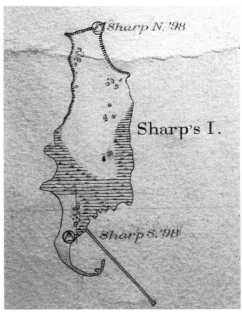

ing on an empty cask. Already fearful of a possible British attack on the town, women and children fled, and the militia prepared for battle. When they discovered their fears were unfounded, the townspeople were slow to forgive Gibson for the scare he'd given them. To make amends, he presented the town with two cannon for defense and thus earned his reprieve.

In the last half of the nineteenth century, Sharps Island was owned by a U.S. army general who served in the Wyoming Territory and a Marquis d'Oyley. Toward the end of the century, Miller R. Creighton of the Baltimore shoe- and boot-manufacturing firm Young, Creighton, and Diggs owned the island. Banking on the great popularity of Victorian-era bayside resorts, Creighton built a hotel on Sharps. The three-story, six-gabled building looked out on a long boardwalk down to the water and a steamboat landing. The Talbot County Assessment records of 1898

listed the value of the hotel as $7,511, an amount that included fifty-four sheep, two mules, a cow, and a boat. Creighton had high hopes to realize a profit, but he failed to take into account the island's rate of erosion. By 1900 the island had shrunk to ninety-four acres, and the steamboat pier was gone. Ultimately, the Sharps Island Hotel was closed and within ten years had disappeared, torn down bit-by-bit, its lumber carried off, as the story goes, to build homes on Tilghman Island.

Beginning in 1830 Sharps Island became an aid to sailors navigating the shoals and ragged edges of the Eastern Bay. That year the federal government built the island's first lighthouse on what was then known as Bay Point. The light stood on a small frame house thirty feet above the ground. Aware of the rapid rate of erosion, the builders constructed the house so that it could be moved if the foundation washed away. Their foresight paid off. By 1848 erosion had whittled the island down to 473 acres and was eating away at the base of the lighthouse. The government's lighthouse board ordered the lighthouse moved inland, but this was little more than a stopgap measure. Bay winds and storms continued to work on the island's soft soil, undermining the bluff on which the lighthouse stood.

The government replaced the original building in 1866 with a screwpile lighthouse, which survived until the winter of 1881 when ice dislodged its supports and carried the house away. Trapped aboard, the keepers rode the lighthouse for sixteen hours until it grounded, and they were rescued along with some furniture and the lens. Part of the island still stood above water in 1882, but the government chose to erect the last light, an automated, caisson-type beacon, on a five-acre underwater plot. Over the ensuing years, the waters of the bay advanced on the remaining land at a remarkable rate of 110 feet a year, and in 1963 Sharps Island disappeared from view. In January 1977 winter's ice left the caisson with a fifteen-degree tilt on its foundation nine feet below the surface. One last major event in the life of the light was replacement of the beacon's glass lens with one of plastic.

In his epic *Chesapeake*, James Michener immortalized Sharps Island, which he used as the model for his fictitious "Devon Island." Today all visible evidence of it is gone. A warning buoy marks the shoal that was once an island, which charts identify simply as the "Sharps Island Obstruction." Some still remember it otherwise. Local waterman Capt. E. Ashton Plummer described it in the way of those whose life has long been tied to the bay, its lands, and waters. "The island was bounded on the north by soft crabs," he said, "on the east by fresh fish, on the south by mosquitoes, and on the west by all three."

REFERENCES

1848 TC, USCGS, 251 (449.29 acres); 1900 TC, USCGS, 2494 (94.18 acres); 1942 Sharps Island Topographic Map (16.52 acres); 1942 TC, USCGS, 8241 (9.55 acres); 1963 HC, USCGS, 8744 (shoal only, no island); 1974 (shoal only); 1998 NOAA 12266 (shoal only). Erosion loss four acres per year.

Bryan, Pat, "The Sharps Island Hotel" in *The Last Hotel: Eastern Shore Summers of a Vanished Way of Life* (Wye Mills, Md.: Chesapeake College Press, 1985). Byron, Gilbert, *The War of 1812 on Chesapeake Bay* (Baltimore: MdHS, 1964). Hornberger/ Turbeyville. Mowbray. Preston, D. J., *Talbot County, A History* (Centreville, Md.: Tidewater, 1983). "Sharps Island Light" (Crownsville, Md.: MdHT, n.d.).

# APPENDIX: LOST ISLANDS

By the end of the twentieth century, hundreds of islands literally had sunk beneath the waves. No longer found on modern charts, they live only in the historical record. More than 500 of them appear below. (The years given refer to dates of purchase or grant.) Because some carried different names, a given island may be listed more than once. References to the islands typically tell us little or nothing about them. Only a few were populated; locals farmed most or simply turned livestock loose on them to graze. On the smallest ones hunters occasionally set up blinds and waited for wildfowl.

Frequently cited sources and their abbreviations:

| | |
|---|---|
| Gelenter | Gelenter, Dave. "Eastern Shore Island Losses." Unpublished report, U. S. Fish and Wildlife Service, 1990. |
| HCaC | Jourdan, Elise. *History of Calvert County, Maryland*. Westminster, Md., 1926. |
| MdSA | Land Records (Patents, Tract Index), 1634. Maryland State Archives, Annapolis, Md., 1985. |
| Nugent | Nugent, Nell Marion. *Cavaliers and Pioneers: Abstracts of Virginia Patents and Grants, 1623. 1800.* Vol. 1 (1623–1666). Vol. 2 (1666–1695). Vol. 3 (1695–1732). Vol. 4 (1732–1741). Vol. 5 (1741–1749). (Richmond: Dietz Printing Co., 1934) |
| USGS | United States Geological Survey |

Alder. Harford County, Md., 1832, 12 acres, 65 perches, Henry Wesley, MdSA.

Alderen. Talbot County, Md., 1695, 41 acres, William Alderen, MdSA.

Angle. Cecil County, Md., 1820, 1 rod, 2 perches, Joseph Miles and Nathaniel Ewing, MdSA.

Augustine Condon's. Cecil County, Md., 1909, acreage unknown, Augustine Condon, MdSA.

Back Creek. Near Tangier Island, Va., 1714, part of 324 acres, Margaret Gibson et al., Nugent, vol. 3.

Bar. Charles County, Md., 1811, 4.5 acres, 18 perches, Nehemiah E. Mason, MdSA.

Barnes. St. Mary's County, Md., 1851, 2 acres, Joshua Barnes, MdSA.

Bass. Charles County, Md., 1904, 25 acres, 2 rods, 25 perches, Richard S. Mitchell, MdSA.

Bear. Anne Arundel County, Md., 1802, Severn River off Bear Branch, between Indian Landing and Hammondswood Landing, 34 sq. perches, William Hammond, MdSA.

Bear. Harford County, Md., 1731, .1 acre, Stephen Onion, MdSA.

Beautiful. Baltimore County, Md., 1741, 62 acres, John Johnson, MdSA.

Beaver. Cecil County, Md., 1767, 4.75 acres, Andrew Barrett, MdSA.

Bee. Dorchester County, Md., 1740, 75 acres, James Hall, MdSA.

Beech. Prince Georges County, Va., 1722, 145 acres above Beech Island, John Scott, Nugent, vol. 3.

Benjamin. Somerset County, Md., 1906, 1 acre, Benjamin Gibson, MdSA.

Benets Park. Date unknown, Isle of Wight County, Va., 700 acres, Goose Hill Creek up to Seawards Creek, Nugent, vol. 1.

Bennetts. Anne Arundel County, Md., 1658, South Creek, 275 acres to Richard Bennett, MdSA.

Bentleys. Dorchester County, Md., 1763, 100 acres, Levin and William Traverse, MdSA.

Big Oyster. Somerset County, Md., 1905, 6 acres, William and B. F. Gibson, MdSA.

Birch. Harford County, Md., 1820, 3 acres, 140 perches, John Kirk, MdSA.

Black. Harford County, Md., 1665, 100 acres, John Collett, MdSA.

Black Duck. Cecil County, Md., 1879, acreage unknown, Henry C. Poplar, MdSA.

Black Duck. Somerset County, Md., 1928, 6.31 acres, Garnett C. Polk et al., MdSA.

Blackhead. Somerset County, Md., 1928, 49.5 sq. rods, Garnett C. Polk, MdSA.

Blackheads. Baltimore County, Md., 1819, 8.25 acres, William Krebs, MdSA.

Blackwalnut. Next to Tangier Island, Va., 1713, 170 acres, John Morris, Nugent, vol. 3.

Bluff. Dorchester County, Md., 1769, 250 acres, William Greene, MdSA.

Bodkin. Harford County, Md., 1819, 45 perches, John Forwood, MdSA.

Bonaparte. Harford County, Md., 1820, 1 acre, 100 perches, Isaac Hawkins, MdSA.

Bonby. Baltimore County, Md., 1819, 1.5 acres, John Uler, MdSA.

Brant. Somerset County, Md., 1925, .72 acre, Levi B. Phillips, MdSA.

Brarary. Somerset County, Md., 1721, 150 acres, John Tunstall, MdSA.

Bread Street. Near Tangier Island, Va., 1711, part of 324 acres, Margaret Gibson et al. Nugent, vol. 3.

Brewers. Anne Arundel County, Md., 1792, in Curtis Creek, 70 perches, Nicholas Brewer, MdSA.

Bridge. Harford County, Md., 1834, 103 perches, John Kirk, MdSA.

Brights. Queen Anne's County, Md., 1746, 28 acres, Francis Bright, MdSA.

Brislan. Harford County, Md., 1815, 15 acres, John Richey, MdSA.

Broken. Harford County, Md., 1706, 133 acres, John Deaver, MdSA.

Brown's. Charles County, Md., 1845, 2.5 acres, Capt. William Brown, MdSA.

Buck. Cecil County, Md., 1825, 1 acre, Thomas Steele et al., MdSA.

Buck. Queen Anne's County, Md., 1762, 145 acres, Richard Everning Harrison, MdSA.

Buck. Queen Anne's County, Md., 1748, 100 acres, Zerobabel Wells, MdSA.

Bullies. Dorchester County, Md., 1896, 365 acres, William R. Leatherbury et al., MdSA.

Burk. Dorchester County, Md., 1759, 145 acres, Richard Harrison, MdSA.

Burntwood. Between Muddy Creek and Messango Creek, Accomac County, Va., 1687, 150 acres, Philip Fisher, Nugent, vol. 2.

Burroes. Lower Norfolk County, Va., 150 acres, Mr. Near, Nugent, vol. 1.

Busters. Dorchester County, Md., 1963, 1.292 acres, F. Leonard Mills, MdSA.

Butchers. At mouth of Taberers Creek, up Hudnalls Creek, Isle of Wight County, Va., 1639, 200 acres, John Davis, Nugent, vol. 1.

Butterfly. Harford County, Md., 1811, 107 perches, John Cline, MdSA.

Buttonwood. Harford County, Md., 1802, 109 perches, John Cline, MdSA.

Buttonwood. Harford County, Md., 1825, 4 acres, 50 perches, John Kirk, MdSA.

Buzzard. Charles County, Md., 1743, 120 acres, William Carter, MdSA.

Buzzards. 1652, Calvert County, Md., 700 acres, William Stone, MdSA.

Cagers. Somerset County, Md., 1664, 200 acres, Robert Cager, MdSA.

Calf. Harford County, Md., 1803, 66 perches, John Ashmore, MdSA.

Campbells. Harford County, Md., 1705, 48 acres, John Campbell, MdSA.

Canady. Somerset County, Md., 1774, 264.5 acres, Isiah Tilghman, MdSA.

Canady. Somerset County, Md., 1809, 5.75 acres, John J. Tull, MdSA.

Canary. Baltimore County, Md., 1963, 3.77 acres, Canary Island Development Company, MdSA.

Canary Island #3. Baltimore County, Md., 1965, 15.805 acres, Canary Island Development Company, MdSA.

Canvasback. Baltimore County, Md., 1826, 2 acres, William Krebs, MdSA.

Canvasback. Somerset County, Md., 1925, 4.24 acres, Levi B. Phillips, MdSA.

Carpenters. St. Mary's County, Md., 1847, 1.75 acres, Susan E. Carpenter et al., MdSA.

Catfish. Baltimore County, Md., 1809, 3.25 acres, Nicholas Forrest, MdSA.

Cedar. South River off Beards Point, Anne Arundel County, Md., 1914, .2 acre, Joseph Bigelow, MdSA.

Cedar. Baltimore County, Md., 1736, 3 acres, John Moale, MdSA.

Cedar. Harford County, Md., 1819, 1 acre, John Kirk, MdSA.

Cedar. Lower Norfolk County, Va., 1674, 74 acres, Robert Symons, Nugent, vol. 2.

Cedar. Lower Norfolk County, Va., 1685, 75 acres, Mathew Godfrey, Nugent, vol. 2.

Cedar. In Great Marsh, Princess Anne County, Va., 402 acres, Evan Jones, Nugent, vol. 3.

Cedar. In Great Marsh between Notts and Main Woods, Princess Anne County, Va., 1711, 402 acres, Devorak and Joseph Godwin, Nugent, vol. 3.

Cedar. Mouth of Nansemund River, Va., date unknown, Nugent, vol. 1.

Cedar. S. side of Warwick River, Va., date unknown, Nugent, vol. 1.

Ceedar. Encompassing island, Mobjack Bay, Va., 920 acres, Nugent, vol. 1.

Chances. Dorchester County, Md., 1918, 248 acres, Daniel J. Willey et al., MdSA.

Channel. Somerset County, Md., 1939, 3.64 acres, H. Harvey Bradshaw, MdSA.

Cherry. Northampton County, Va., 1823, part of 2,800 acres on Arracocke Creek and Nandua Creek, Capt. Southy Littleton, Nugent, vol. 2.

Chinch. Dorchester County, Md., 1923, 1 acre, 2 rods, 27 perches, Albanus Phelps, MdSA.

Chinkapin. S. side of Round-and-Out Bridge, Henrico County, Va., 1702, 51 acres, Capt. Wm. Sloanes, Nugent, vol. 3.

Clover. Harford County, Md., 1819, 30 perches, John Kirk, MdSA.

Clows. Queen Anne County, Md., 1763, 11.5 acres, James Clow, MdSA.

Coartneys. Dorchester County, Md., 1755, 225 acres, William Hayne, MdSA.

Cobhams. On Pocomoke River near Back Creek, near Mollies Neck, Accomac County, Va., 100 acres, Maj. Gen. Jno. Custis, Nugent, vol. 2.

Cockeys. Off Eastern Neck Island, Kent County, Md., 5.7 acres, Josh Cockey, Gelenter.

Conjurers. Dorchester County, Md., 1667, 3 acres, Thomas Hooten, MdSA.

Cooks. Harford County, Md., 1770, 5 acres, Robert Cook, MdSA.

Copper. Harford County, Md., 1806, .5 acre, 17 perches, Capt. Thomas McKimmon, MdSA.

Corsica. Charles County, Md., 1776, 1.25 acre, William Smallwood, MdSA.

Cortneys. Somerset County, Md., 1667, 100 acres, Thomas Cortney, MdSA.

Costens. Mouth of Hickory Null Creek in Elizabeth River, Lower Norfolk County, Va., 1832, Walter Costen, Nugent, vol. 2.

Cotton. Severn River, W. of Indian Landing, Anne Arundel County, Md., 1908, 25.75 perches, Jane Cotton, MdSA.

Cowells. Dorchester County, Md., 1892, 2 acres, 107 perches, Luther Phillips, MdSA.

Cows, location and date unknown, 6.2 acres, Gelenter.

Coxes. Lancaster County, Va., 1653, 1,150 acres, John Cox, Nugent, vol. 1.

Crab. Dorchester County, Md., 1658, 50 acres, Jonathan Batt, MdSA.

Cragbourne. St. Mary's County, Md., 1745, 13 acres, John Chesley, MdSA.

Crane. Charles County, Md., date unknown, 29 perches, Pearson Chapman, MdSA.

Crane. Talbot County, Md., 1787, 11.75 acres, Samuel Barrow, MdSA.

Craney. Harford County, Md., 1814, 2 acres, 3 rods, 15 perches, Polaske Sweeney et al., MdSA.

Craney. Mouth of Elizabeth River, Lower Norfolk County, Va., 50 acres, no owner mentioned, Nugent, vol. 1.

Crow. Lower Norfolk County, Va., 1682, in Currituck Bay near Kings Point, 900 acres, Patrick White, Nugent, vol. 2.

Crowe. Dorchester County, Md., 1701, 35 acres, Attaway Pattison, MdSA.

Crows. Essex County, Va., 1705, 57 acres, Capt. Edward Gouldman, Nugent, vol. 2.

Cutgap. Somerset County, Md., 1871, 1.75 acres, William Whereton, MdSA.

Delfes. Harford County, Md., 1667, 155 acres, Henry Stockett, MdSA.

Delph. Harford County, Md., 1750, 375 acres, Thomas White, MdSA.

Demi-John. Queen Anne's County, Md., 1962, .69 acre, Walter Heed, MdSA.

Desirable. Harford County, Md., 1802, 32 perches, John Sample, MdSA.

Dinner. Harford County, Md., 1802, 86 perches, John Sample, MdSA.

Doegs. Westmoreland County, Va., 1694, 3,609 acres, includes Mylampses and Miopses Islands, William Parker, Nugent, vol. 1.

Doggs. Northumberland County, Va., 1651, 2,109 acres, Richard Turney, Nugent, vol. 1.

Double. Cecil County, Md., 1818, 1 acre, Joseph Miles, MdSA.

Douglas. St. Mary's County, Md., 1957, 1 acre, Otis W. Douglas, MdSA.

Duck. Charles County, Md., 1904, 10 acres, 1 rod, 39 perches, Richard S. Mitchel, MdSA.

Duck. Dorchester County, Md., 144 acres, Milbourne J. Elliott, MdSA.

Duck. Talbot County, Md., 1916, 52 acres, 2 rods, 2 perches, John H. Mullikin et al., MdSA.

Dumocks. Dorchester County, Md., 1904, 1 acre, 1 rod, 7 perches, John B. Dumock, MdSA.

Dumplings. Nansemond County, Va., date unknown, 350 acres beginning at Nansemond River, William Parker, Nugent, vol. 1.

Dumplinns. Nansemond County, Va., 1648, 150 acres, Leonard Gwinns, Nugent, vol. 1.

Earicksons. Queen Anne's County, Md., 1663, 20 acres, Mathew Earickson, MdSA.

Eastern. Kent County, Md., 1662, 100 acres, John Meconokin, MdSA.

East Troy. Somerset County, Md., 1959, 1.68 acres, H. Harvey Bradshaw, MdSA.

Edels #1. Harford County, Md., 1903, 68 acres, 21 perches, Samuel T. Edel et al., MdSA.

Edwin Warfield. Cecil County, Md., 1905, 2.5 acres, John Stump et al., MdSA.

Ege. Va., 1705, 100 acres, Devorax and Joseph Godwin, Nugent, vol. 3.

Ellises. Charles County, Md., 1845, 11.25 acres, Samuel and William Marders, MdSA.

Fabs. Dorchester County, Md., 1964, 2.754 acres, F. Leland Miles, MdSA.

Featherstones. S. side of Rappahannock River at upper end of Portobacco Bay, Va., 37 acres, Henry Beverly, Nugent, vol. 3.

Federal. Cecil County, Md., 1802, .25 acre, John Conrad et al., MdSA.

Fiddlers. Dorchester County, Md., 1900, 15 acres, Levi D. Traverse, MdSA.

First Pier. Harford County, Md., 1836, 11.25 acres, John Kirk et al., MdSA.

Fishbone. Herne Straits, Va., 18 acres, Francis Mackenny et al., Nugent, vol. 3.

Fishcove. Harford County, Md., 1815, 24.25 acres, Col. John Dick et al., MdSA.

Fisher. Somerset County, Md., 1684, 20 acres, George Cave, MdSA.

Flat. Rhode River, Anne Arundel County, Md., 1922, 1.35 acres, Herman Quade and Wm. Bergman, MdSA.

Flea. S. side Appomattox River, Elizabeth City County, Va., 1653, 1,557 acres, part of grant to Maj. Abraham Wood, Nugent, vol. 1.

Flowers. S. side of Chickahominy River, Va., date unknown, 174 acres, Joseph Bradley, Nugent, vol. 3.

Fortunate. Harford County, Md., 1815, 94 acres, John Ashmore, MdSA.

Fortunate. Harford County, Md., 1818, 5.25 acres, John Forwood, MdSA.

Fox. Accomac County, Va., 1678, 83 acres, Thomas Wellborn, Nugent, vol. 2.

Fox. Somerset County, Md., 1930, 1 acre, George Myers, MdSA.

Frederick Stump. Cecil County, Md., 1905, 1 acre, 3 rods, 4 perches, John Stump et al., MdSA.

Friendly. Cecil County, Md., 1798, 2.75 acres, John Archer and Philip Thomas, MdSA.

Friendly. Harford County, Md., 1814, 148 perches, John Picks, MdSA.

Friendly. Harford County, Md., 1838, 49 acres, 1 rod, 6 perches, John Kirk, MdSA.

Gabriels. Near Watts Island (Little Watts and Goats Islands),Va., date unknown, 24 acres, Francis Mackenny and Henry Jenkins, Nugent, vol. 3.

Gap. Harford County, Md., 1856, 3 acres, 1 rod, 7 perches, John Kirk et al., MdSA.

Glass House. James County, Va., 1654, 24 acres, Maj. Francis Morrison, Nugent, vol. 1.

Goates. Includes island in mouth of Rappahannock River, Lancaster County, Va., date unknown, 800 acres, Mrs. Elinor Brocas, Nugent, vol. 1.

Golden. Harford County, Md., 1803, 4.5 acres, 12 perches, John Sample, MdSA.

Goldsborough. Talbot County, Md., 1827, 11.75 acres, Greenbury Goldsborough, MdSA.

Goodins (or Goddins). On Pamunkey River, Kent County, Va., part of 918 acres, Marke Warkeman, Nugent, vol. 2.

Goose. Harford County, Md., 1814, 112 acres, John Riutey, MdSA.

Goose. Harford County, Md., 1833, 3 acres, James and Thomas Fisher Jr., MdSA.

Goose. Somerset County, Md., 1925, 8.36 acres, Levi B. Phillips, MdSA.

Grass. Harford County, Md., 1902, 28 perches, John Sample, MdSA.

Great. Magothy River, opposite Zachariah Gray, next to Raspberry Island, Anne Arundel County, Md., 1769, 12 acres to Wm. Gambrill, MdSA.

Great and Little. James City County, Va., 1650, 239 acres (Great), 89 acres (Little), Edward Sanderson, Nugent, vol. 2.

Greenpoint. Somerset County, Md., 1892, 4.5 acres, William Ellinger, MdSA.

Guys. Somerset County, Md., 1915, .15 acre, Guy Sterling, MdSA.

Halfmoon. On S. side Hunting Creek, Northampton County, Va., 1671, 100 acres, Richard Hill, Nugent, vol. 2.

Halfway. Harford County, Md., 1836, 111 perches, Jeremiah Haines and John Kirk, MdSA.

Hambleton. Two mi. SW of St. Michaels, Talbot County, Md., 85 acres, Gelenter.

Hanging, Calvert County, Md., St. Leonard's Creek, HCaC.

Harbor. Talbot County, Md., 1804, 43.25 acres, Charles Carroll, MdSA.

Harrison. Charles County, Md., 1851, 300 acres, Thomas Harris, MdSA.

Haskins. Cecil County, Md., 1818, 2 rods, 25 perches, Isaac Haskins, MdSA.

Hawkins. Harford County, Md., 1815, 38 perches, Isaac Hawkins, MdSA.

Head. Cecil County, Md., 1816, 2 acres, 6 rods, 38 perches, Reuben Reynolds, MdSA.

Henn. Kent County, Md., 1743, 4 acres, Simon Wilmer, MdSA.

Henrico. Henrico County, Va., 1666, part of 1,000 acres, Thomas Boswell, Nugent, vol. 1.

Henricus. Henrico County, Va., 1619, less than 100 acres, John Leyden, Nugent, vol. 1.

Henry Condons. Cecil County, Md., 1909, 3 acres, 1 rod, 1 perch, Clorita Condon, MdSA.

Henry Stump. Cecil County, Md., 1905, 3 rods, 20 perches, John Stump et al., MdSA.

Herb. Dorchester County, Md., 1962, .689 acre, Herbert H. Tyler, MdSA.

Hernre. Talbot County, Md., 1685, 75 acres, Edward Barracliff, MdSA.

Heron. St. Mary's County, Md., 1929, 3.99 perches, Lewis F. Abel, MdSA.

Herrig. Cecil County, Md., 1820, 2 acres, 2 poles, 2 perches, Joseph Miles et al., MdSA.

Herring. Harford County, Md., 1802, 27 perches, John Ashmore, MdSA.

Herring. Harford County, Md., 1857, 25 acres, 2 rods, 35 perches, John Kirk et al., MdSA.

Herring. Talbot County, Md., 1847, 1 acre, T223.

Herring. Talbot County, Md., date unknown, acreage unknown, on Bennetts Point, Gelenter.

Hoags. Dorchester County, Md., 1710, 53 acres, Charles Mackeel, MdSA.

Hog. Charles County, Md., 1770, 80 acres, Thomas Owings, MdSA.

Hog. Dorchester County, Md., 1716, 50 acres, Lewis Griffith, MdSA.

Hog (part of). James City County, Va., 1634, 50 acres, Abraham Roote, Nugent, vol. 1.

Hog (part of). James City County, Va., 1636, John Dunston, Nugent, vol. 1.

Hog (all of). James City County, Va., date unknown, Randall Holt, Nugent, vol. 1.

Hog. Up Little Creek, Lower Norfolk County, Va., date unknown, Nicholas Huggins, Nugent, vol. 2.

Hog. Queen Anne's County, Md., 1926, 2.5 acres, Tilghman Eaton et al., MdSA.

Hog. St Mary's County, Md., 1821, 6.25 acres, Joseph Stone, MdSA.

Hog. Surrey County, Va., date unknown, 1,450 acres, Randall Holt, Nugent, vol. 2.

Hogg. Charles County, Md., 1734, 460 acres, Charles Musgrove, MdSA.

Hogg. Dorchester County, Md., 1666, 300 acres, Thomas Powell, MdSA.

Hogg. James City County, Va., 1644, 500 acres, James Taylor and Lawrence Baker, Nugent, vol. 1.

Hogg. James City County, Va., 1664, 50 acres (3 small hammocks joined by "ostrums" to Hog Island), Charles Edgerton, Nugent, vol. 1.

Hoggs. Charles County, Md., 1714, 141 acres, John Harrison, MdSA.

Hoggs. Talbot County, Md., 1705, 140 acres, John Keld, MdSA.

Hog Ruting. Dorchester County, Md., 1795, 6.75 acres, Thomas Breenwood, MdSA.

Honey. Harford County, Md., 1814, 12 acres, George McCausland and John Richey, MdSA.

Honney. Cecil County, Md., 1724, 39 acres, Edmund Perke, MdSA.

Holland. Dorchester County, Md., 1818, 91.5 acres, Nathan Todd et al., MdSA.

Hope (also Morgans). James City County, Va., 1650, 107 acres, Edward Sanderson, Nugent, vol. 2.

Hopkins. Nanticoke River, Md., year unknown, Bloodsworth Island topographical map, USGS.

Horney Point. Near Tangier Island, Va., 1843, part of 324 acres, Margaret Gibson et al., Nugent, vol. 3.

Huckleberry. Anne Arundel County, Md., 1743, 30 acres, Isaac Hall, MdSA.

Hughes. Harford County, Md., 1673, 50 acres, Thomas Heath, MdSA.

Imolys. Talbot County, Md., 1850, 25 acres, 21 rods, 23 perches, Edward Goccorthum, MdSA.

Indian. Harford County, Md., 1820, 4.75 acres, John Kirk, MdSA.

Inner. Somerset County, Md., 1943, 43 rods, M. Morris Whitehurst, MdSA.

Island. Dorchester County, Md., 1667, 35 acres, John Puccard, MdSA.

Island #1. Dorchester County, Md., 1929, 240 sq. feet, William H. Francis, MdSA.

Island #2. Dorchester County, Md., 1929, 23.8 perches, William H. Francis, MdSA.

Island #3. Dorchester County, Md., 1925, 2 acres, 52 perches, Edward J. Phillips, MdSA.

Island #4. Dorchester County, Md., 1925, 8 9/40 acres, Edward J. Phillips et al., MdSA.

Island #5. Dorchester County, Md., 1925, 2 acres, 52 perches, Edward J. Phillips et al., MdSA.

Island, The. In Severn River, off Brooksbys Point, Anne Arundel County, Md., 1772, 1 acre, John Hammond, MdSA.

Island, The. Charles County, Md., 1792, 14.25 acres, George Tubman, MdSA.

Island, The. Dorchester County, Md., 1681, 50 acres, Bartholomew Ennals, MdSA.

Island of Adamant. Harford County, Md., 1802, 63 acres, Cooper Boyd, MdSA.

Island of Caparee. Harford County, Md., 1798, 50.5 acres, John Cockey, MdSA.

Island of Capria. Harford County, Md., 1819, 72.5 acres, George McCausland, MdSA.

Island of Little Worth. Anne Arundel County, Md., 1813, 1 acre, 22 perches, James Corneill, MdSA.

Island of Malta. Harford County, Md., 1820, 1.5 acres, Isaac Hawkins, MdSA.

Island of Man. Harford County, Md., 1805, 76 perches, John Sample, MdSA.

Island of May. Harford County, Md., 1818, 12.5 acres, George McCausland, MdSA.

Island of St. Helena. Dorchester County, Md., 1938, 2 acres, 2 rods, 29 perches, Isaac Waller, MdSA.

Island of St. Johns. Harford County, Md., 1814, 89 perches, John Richey, MdSA.

Island of Sky. Harford County, Md., 1806, .75 acre, Thomas McKinnon, MdSA.

Island of Wight. Dorchester County, Md., 1683, 50 acres, William Arundel, MdSA.

Isle of Pines. Cecil County, Md., 1837, 18.5 acres, Jonas Person Jr., MdSA.

James. James City County, Va., 1632, 250 acres, Wm. Spencer, Nugent, vol. 1.

James. James City County, Va., 1637, 1 acre, Alexander Stomar, Nugent, vol. 1.

James. James City County, Va., 1637, 6 acres, John Corker, Nugent, vol. 1.

James. James City County, Va., 1637, 12 acres, John Radish and John Bradwell, Nugent, vol. 1.

James. James City County, Va., 1647, 300 acres, Wm. Lawrence, Nugent, vol. 1.

Jerrys. Harford County, Md., 1856, 4 acres, 23 perches, John Kirk, MdSA.

Joe Scarborough's. Harford County, Md., 1847, 17 acres, John Carrol Walsh, MdSA.

Johnson's. Harford County, Md., 1702, 60 acres, Daniel Johnson, MdSA.

Johnson's. Cecil County, Md., 1796, 28.75 acres, John and Nathaniel Kerr, MdSA.

John Stump. Cecil County, Md., 1905, 3 rods, 13 perches, John Stump et al., MdSA.

Jones. Somerset County, Md., 1671, 100 acres, Leonard Jones, MdSA.

Jordans. Chester River, Md.

Josephs. E. side Custices Creek, Anne Arundel County, Md., 1748, 9 acres, Thomas Hall, MdSA.

Kagers. Somerset County, Md., 1665, 200 acres, Robert Kager, MdSA.

Kennedeys. Toward Rich Island S. to Chuckatuck River, Lower Norfolk County, Va., 1638, 50 acres, Tristum Nosworthy, Nugent, vol. 1.

Kerrs. Cecil County, Md., 1787, 5.25 acres, Nathaniel and John Kerr, MdSA.

Kerrs. Harford County, Md., 1814, 1 acre, 64 perches, James Kerr, MdSA.

Kidds. Anne Arundel County, Md., 1904, 20 acres, Robert Welsh, MdSA.

Knots. Currituck Bay, Lower Norfolk County, Va., 1682, 327 acres, Patrick White, Nugent, vol. 2.

Landing. Dorchester County, Md., 1889, 128 acres, 1 rod, 14 perches, Elijah Hurley, MdSA.

Lawsons. S. side Occupason Creek, Va., 1665, Nugent, vol. 1.

Lewis. Divided by River and Pounces Creek on S., New Kent County, Va., 1,000 acres, John Lewis and James Turner, Nugent, vol. 1.

Little. Dorchester County, Md., 1764, 18 acres, Anne Griffith, MdSA.

Little. Dorchester County, Md., 1769, 48 acres, Anne Griffith, MdSA.

Little. Dorchester County, Md., 1926, 11 sq. perches, Lasbury Parks et al., MdSA.

Little McDowell. Cecil County, Md., 1905, 1 rod, 2 perches, Frederick Irwin, MdSA.

Little Oyster. Somerset County, Md., 1905, 1 acre, 2 rods, William E. and Benjamin Gibson, MdSA.

Long. Beginning at Nantecock S. to Great Bay of Chesapeake, Accomac County, Va., 1743, 18 acres, Francis Mackemie et al., Nugent, vol. 3.

Long. Cecil County, Md., 1820, 10 acres, 28 perches, John Miles and Nathaniel Ewing, MdSA.

Long. One mi. NW of Holland Island, Dorchester County, Md., unknown acreage, Gelenter.

Long. Harford County, Md., 1802, 120 perches, John Sample, MdSA.

Long. Harford County, Md., 1819, 31 perches, John Kirk, MdSA.

Long. Harford County, Md., 1824, 16.5 acres, Christian Hoffman, MdSA.

Long. Lower Norfolk County, Va., 1674, Currituck Bay, 250 acres, Peter Malbone, Nugent, vol. 2.

Long. In Back Bay of Corotock, Princess Anne County, Va., 1711, 294 acres, John Molbone, Nugent, vol. 3.

Long Marsh. Queen Anne's County, Md., 1929, 10.55 acres, G. H. Metcalfe, MdSA.

Long Pond. Isle of Wight County, Va., 1637, 250 acres, Wm. Eyres, Nugent, vol. 1.

Lord. Henrico County, Va., 1637, 20 acres, Arthur Bailey, Nugent, vol. 1.

Lottery. Harford County, Md., 26 acres, Stephen Fisher, MdSA.

Maddox. Somerset County, Md., 1960, .35 acre, H. Harvey Bradshaw, MdSA.

Mads. Selby Bay, Anne Arundel County, Md., 1809, 3.125 acres, Nicholas Forrest, MdSA.

Mallard. Somerset County, Md., 1928, 1.2 acres, Garnet P. Pock et al., MdSA.

Manlone. Somerset County, Md., 1905, 2.5 acres, James Graham and F. Bradshaw, MdSA.

Many Islands. Harford County, Md., 1836, 80 acres, Henry Weslew and Henry Moore, MdSA.

Maple. Cecil County, Md., 1903, 1 rod, 9 perches, Frederick Irwin, MdSA.

Marsh. Dorchester County, Md., 1723, 196 acres, William Greene Sr., MdSA.

Marsh. Dorchester County, Md., 1803, 43 acres, Levin Lewis, MdSA.

McDormans. Somerset County, Md., 1853, 4 acres, William H. McDorman, MdSA.

McVeys. Cecil County, Md., 1913, 10 acres, Joseph Condon, MdSA.

Merry Bayle. Rappahannock County, Va., 1658, included in 450 acres, Robert Younge, Nugent, vol. 1.

Middle. Dorchester County, Md., 1793, 53.5 acres, William Craswell, MdSA.

Miles. Charles County, Md., 1761, 104.25 acres, Edward Miles, MdSA.

Milkum. Cecil County, Md., 1714, 1,000 acres, Mathew VanBeeber, MdSA.

Miltons. Somerset County, Md., 1943, 14 rods, M. M. Whitehurst, MdSA.

Minimum. Somerset County, Md., 1902, 3 acres, 3 rods, 21 perches, William F. McDorman, MdSA.

Mink. Harford County, Md., 1818, 42 perches, John Kirk, MdSA.

Mitchells. Kent County, Md., 1807, 1 acre, 2 rods, 3.75 perches, Richard B. Mitchell, MdSA.

Monocky. Harford County, Md., 1888, 1 acre, 29 perches, Hannah and Richard DeCon, MdSA.

Morgans. Several small islands near James City County, Va., 1750, Edward Sanderson, Nugent, vol. 3.

Morses. Anne Arundel County, Md., 1737, 25 acres, Richard Morse, MdSA.

Mulberry. Dorchester County, Md., 1658, 50 acres, Robert Wilson, MdSA.

Mulberry. Dorchester County, Md., 1749, 20.5 acres, Andrew Insley Jr., MdSA.

Mulberry. Bakers Neck, James River, Va., 1619, part of grant up bank of Warwick River, Capt. William Pierce, Nugent, vol. 1.

Mulberry. Head of Kethes Creek, James City County, Va., 1665, 100 acres, Richard Adkins, Nugent, vol. 1.

Mulberry. James City County, Va., 1667, Willis Helly, Nugent, vol. 1.

Mulberry. Warwick County, Va., 1669, 1,350 acres, Theo Iken, Nugent, vol. 2.

Murders. Charles County, Md., 1846, 5 acres, 2 rods, 20 perches, Thomas and Wm. Murders, MdSA.

Muskrat. Anne Arundel County, Md., 1818, 3 acres, 37 perches, Benjamin McCeney, MdSA.

Muskrat. Harford County, Md., 1802, 24 perches, John Ashmore, MdSA.

Mussell. Westmoreland County, Va., 1657, part of 3,000 acres at head of Niopsic Creek, NE upon Ochaquim River, no owner mentioned, Nugent, vol. 1.

Natts. Harford County, Md., 1658, 100 acres, Thomas Overton, MdSA.

Near. Harford County, Md., 1795, 23.5 acres, John Street, MdSA.

New Point Comfort. At mouth of creek running into Dyers Creek, Gloucester County, Va., 1684, 300 acres, Edward Davis, Nugent, vol. 3.

Northeast. Dorchester County, Md., 1903, 36.25 acres, Isaac White et al., MdSA.

Notts. Princess Anne County, Va., 1711, part of 402 acres in Great Marsh on uppermost fork of Cornew Creek to Back Bay of Corotuck, Evan Jones, Nugent, vol. 3.

Oake. Northampton County, Va., 3,600 acres includes two hummocks, Oake Island and Pond Island, William Whittington, Nugent, vol. 2.

Oaken. Through Cedar Island marshes to Oaken Island, Mobjack Bay, Va., 1632, part of 1,500 acres, Nugent, vol. 1.

Old Walstones. Lower Norfolk County, Va., 50 acres, now Piney Island, James Pewters, Nugent, vol. 2.

Olivias. Somerset County, Md., 1906, 1 acre, Olivia M. Gibson, MdSA.

Otter. Cecil County, Md., 1817, 1 acre, 1 rod, 13 perches, Reuben Reynolds, MdSA.

Otter. Harford County, Md., 1802, 95 acres, John Ashmore, MdSA.

Otter. Harford County, Md., 1814, 14 perches, John Dicke, MdSA.

Otter. Harford County, Md., 1811, .75 acre, 14 perches, Robert Amos et al., MdSA.

Otter. Somerset County, Md., 1960, .25 acre, Clinton W. Corbin, MdSA.

Ox and Bird Islands. On Chickahominy River S. of Hookers Mill Creek, James City County, Va., 1728, 37 acres, Philip Smith, Nugent, vol. 3.

Oyster, Anne Arundel County, Md., 1912, .15 acre, Ernest Hall and Gladys Bousch, MdSA.

Oyster. Talbot County, Md., 1874, 4 perches, William Martin, MdSA.

Oystershell. Anne Arundel County, Md., 2 acres, John Ridout, MdSA.

Papa. Dorchester County, Md., 1665, 50 acres, William Smith, MdSA.

Papacoone. NE side of Chickahominy River at Richahoc Neck, Virginia, 1656, part of 2,000 acres, Henry Sloane, Nugent, vol. 1.

Parrott. Harford County, Md., 1818, 4 perches, John Kirks, MdSA.

Parsimmon. Harford County, Md., 1814, 53 acres, Isaac Hawkins, MdSA.

Patience. Lower Norfolk County, Va., 1671, 100 acres, part being Patience (Woolstones Island), Thomas Woolstone, Nugent, vol. 2.

Pauls. Harford County, Md., 1803, 86 perches, Cooper Boyd, MdSA.

Pelew. Harford County, Md., 1804, 65 perches, John Ashmore, MdSA.

Perks's. Cecil County, Md., 1728, 81 acres, Stephen Onion, MdSA.

Peverly. Harford County, Md., 1820, 40.44 acres, Sidney Peverly, MdSA.

Peverly #2. Harford County, Md., 1927, 1.58 acres, Sidney Peverly, MdSA.

Peverly #3. Harford County, Md., 1927, 3.12 acres, Sidney Peverly, MdSA.

Phillips. W. side Chickahominy River, Va., 1653, Thomas Stampe, Nugent, vol. 1.

Pine. Gloucester County, Va., 1689, 5 islands, North River, Augustine Horth, Nugent, vol. 2.

Piney. Lower Norfolk County, Va., 1674, 250 acres, includes Piney Island in Surry Plantation, Thomas Griffin, Nugent, vol. 2.

Piney. Somerset County, Md., 1723, 10 acres, Robert King, MdSA.

Pink. Harford County, Md., 1819, 1 perch, John Kirk, MdSA.

Pitcraft. Somerset County, Md., 1960, .21 acre, Harvey Bradshaw, MdSA.

Pleasants. Appomattox River, Va., 1722, 16 acres includes 5 islands, Nicholas Overby, Nugent, vol. 3.

Plover. Somerset County, Md., 1925, .73 acre, Levi B. Phillips, MdSA.

Pluvers. Anne Arundel County, Md., 1860, 3 acres, 8 perches, Elijah Stalling, MdSA.

Point. Somerset County, Md., 1929, 53.62 sq. poles, M. M. Whitchurst, MdSA.

Point Comfort. Elizabeth City County, Va., 1627, 50 acres, James Bunall, Nugent, vol. 1.

Point Comfort. Elizabeth City County, Va., 1638, 100 acres, William Armistead, Nugent, vol. 1.

Poles. Ely County, Va., 1638, 24 poles square, William Parry, Nugent, vol. 1.

Polters. Princess Anne County, Va., 1711, 85 acres, 5 islands, Henry Spratt, Nugent, vol. 3.

Pond. Northampton County, Va., 1609, 3,600 acres includes two hummocks, Oake Island and Pond Island, William Whittington, Nugent, vol. 2.

Pooles. Warwick County, Va., 1643, part of 550 acres (Warwick River), Alice Harlow, Nugent, vol. 2.

Possimon. On James River, Charles City County, Va., 1715, 28 acres, Nathaniel Morrison, Nugent, vol. 3.

Powells. End of Howells Point, Choptank River, Md., 1755, 55 acres, Judge Samuel Dickinson, Gelenter.

Powells. Talbot County, Md., 1666, 50 acres, Howell Powell, MdSA.

Powhite. Va., 1690, 780 acres, Jeremiah Benskin, Nugent, vol. 2.

Pratts. St. Mary's County, Md., 1819, 3.5 acres, James R. Pratt, MdSA.

Prickley Pair. Anne Arundel County, Md., 1738, 50 acres, Clement Mattingley, MdSA.

Pry. Somerset County, Md., 1925, 10.9 acres, Earnest J. Gore and Amos Phillips, MdSA.

Punch. Dorchester County, Md., 1.3 acres, Gelenter.

Rabbit. Anne Arundel County, Md., 1747, 2.5 acres, Nicholas MacCubbin, MdSA.

Racoune. Gloucester County, Va., 1660, 100 acres, Thomas Boswell, Nugent, vol. 1.

Racoune. Va., 1705, 100 acres, N. point of Ware River, George Jansen Nugent, vol. 1.

Ragged. Lower Norfolk County, Va., 1690, 50 acres adjacent to Cedar Island, Joseph Percy, Nugent, vol. 2.

Ragged. Upper Norfolk County, Va., 1635, 100 acres, John Seaward, Nugent, vol. 1.

Ragged. Upper Norfolk County, Va., 1639, islands behind Poplar Neck, Rich Island, Canada Island, Kennedyes Island, 100 acres, Tristram Nosworthy, Nugent, vol. 1.

Rattlesnake. Dorchester County, Md., 1722, 250 acres, William Ennolls, MdSA.

Rattlesnake. Dorchester County, Md., 1722, 280 acres, William Ennolls, MdSA.

Red. Dorchester County, Md., 1753, 45 acres, Thomas Manning, MdSA.

Redhead. Somerset County, Md., 1928, 34.25 sq. rods, Garnett C. Polk et al., MdSA.

Reed Bird. Anne Arundel County, Md., 1909, 33.75 acres, John P. Bruns, MdSA.

Rees. Harford County, Md., 1819, 6 perches, John Kirk, MdSA.

Reuben. Cecil County, Md., 1903, 2 acres, 1 pole, 25 perches, Frederick Irwin, MdSA.

Rich. Dorchester County, Md., 1770, 14 acres, Thomas White, MdSA.

Rich Island Addition. Dorchester County, Md., 1773, 24 acres, Thomas White, MdSA.

Riggins. Somerset County, Md., 1856, 7.875 acres, Severn Riggin, MdSA.

Rock. Cecil County, Md., 1820, 1 acre, 3 poles, 28 perches, Joseph Miles and Nathaniel Ewing, MdSA.

Rock. Cecil County, Md., 1860, 70 perches, Jethro Johnson, MdSA.

Rock. Somerset County, Md., 1943, 1.5 acres, M. Morris Whitehurst, MdSA.

Rockay. Harford County, Md., 1818, 20 perches, John Kirk, MdSA.

Rockes. Harford County, Md., 1814, 63 perches, John Dicks, MdSA.

Rockford. Cecil County, Md., 1818, 38 perches, James Ewing, MdSA.

Rocky. Harford County, Md., 1796, 2 acres, Henry Stump, MdSA.

Rogues. In Piscataway Creek, Essex County, Va., 1697, 80 acres, Jonathan Fisher, Nugent, vol. 3.

Rogues. Beginning at Perrys Creek, Va., 1730, 80 acres, Henry Pickett, Nugent, vol. 3.

Rosemary Lane. Tangier Island, Va., 1714, near part of 324 acres, Margaret Gibson et al., Nugent, vol. 3.

Ross's. Dorchester County, Md., 1781, 32 acres, Anthony Ross, MdSA.

Rough. Cecil County, Md., 1817, .75 acre, 20 perches, John Kirk, MdSA.

Royston's (formerly Alderns). One mi. SW of mouth of Irish Creek, Md., 1755, 41 acres, Elizabeth Aldern, Gelenter.

Ruark's. Dorchester County, Md., 1799, 542.5 acres, James Ruark, MdSA.

Ruths. Queen Anne's County, Md., 1765, 2.5 acres, William Ruth, MdSA.

Sampsons. Lower Norfolk County, Va., 3,100 acres includes Sampsons Island at mouth of Northwest River, John Gibbs, Nugent, vol. 2.

Sams Bush. Somerset County, Md., 1871, 9.75 acres, Cornelius and Aaron Sterling, MdSA.

San Domingo. Harford County, Md., 1834, 2 acres, John Kirk et al., MdSA.

Sand Pier. Somerset County, Md., 1904, 4.5 acres, Robert H. Jones Jr., MdSA.

Sandy. Northampton County, Va., 90 acres, Thomas Moore, Nugent, vol. 2.

Sandy. Surry County, Va., 1714, part of 90 acres bounded by Hog Island on S., Revells Island on E., marshes on N.; William and Nathaniel Ball, Nugent, vol. 3.

Sandy Beach. Next to Tangier Island, Va., 1713, 170 acres, John Morris, Nugent, vol. 3.

Sandy Point. Cecil County, Md., 1903, 14 acres, Frederick Irwin, MdSA.

Scarburgh's Winter. N. side of Pungoteague Creek, Accomac County, Va., 1687, 30 acres, Maj. Charles Scarburgh, Nugent, vol. 2.

Scaup. Somerset County, Md., 1925, .34 acre, Levi P. Phillips, MdSA.

Second Pier. Harford County, Md., 1856, 10.25 acres, John Kirk et al., MdSA.

Secrets. Anne Arundel County, Md., 1805, 126 sq. perches, Richard Harwood, MdSA.

Security. Cecil County, Md., 1814, 8 acres, 2 rods, 26 perches, Hugh Steel, MdSA.

Sedge. Harford County, Md., 1832, 1 acre, 3 rods, 2 perches, Joseph Miles, MdSA.

Seven. On Fluvanna River, Goochland County, Va., 175 acres, Thomas Tindal, Nugent, vol. 3.

Shad. Cecil County, Md., 1814, 135 acres, 2 rods, 8 perches, Samuel Kerr, MdSA.

Shad. Harford County, Md., 1814, 102 perches, Isaac Hawkins, MdSA.

Shad. Harford County, Md., 1818, 90 perches, John Kirk, MdSA.

Shad. Harford County, Md., 1818, 39.75 acres, Thomas Pattmeld, MdSA.

Sheep. Dorchester County, Md., 1924, 15 acres, 3 rods, 15 perches, Tavinia J. Todd, MdSA.

Sherand. Harford County, Md., 1808, 5 acres, 3 rods, 14 perches, John Sample, MdSA.

Sherwoods. Talbot County, Md., 1681, 20 acres, Miles River, Hugh Sherwood, MdSA, Gelenter.

Shetland. Cecil County, Md., 1833, 10 acres, Alexander E. Grubb, MdSA.

Small. Harford County, Md., 1795, 2 acres, John Sterrt, MdSA.

Smallwoods. Charles County, Md., 1761, 18.25 acres, Bayne Smallwood, MdSA.

Smiths. Cecil County, Md., 1903, 1 acre, Frederick W. Irwin, MdSA.

Smiths. N. side of Rappahannock River, on N. side of Robinson River, Spotsylvania County, Va., 1789, 196 acres, George Woods, Nugent, vol. 3.

Snakes. Harford County, Md., 1794, 2 acres, Nathaniel Smith et al., MdSA.

Spicers. Anne Arundel County, Md., 1725, 3 acres, John Grace, MdSA.

Squirrel. Harford County, Md., 1823, 9 acres, 34 perches, John Kirk, MdSA.

Stacey. Harford County, Md., 2.5 acres, 20 perches, John Evat, MdSA.

Steels. Harford County, Md., 1819, 24 perches, John Kirk, MdSA.

St. Stephens. Harford County, Md., 1818, 13 perches, John Forwood, MdSA.

Stevens. Talbot County, Md., 1768, 45 acres, William Stevens, MdSA.

Stineman's. Cecil County, Md., 1948, 6.52 acres, John Steinman et al., MdSA.

St. Johns. Somerset County, Md., 1668, 80 acres, Thomas Taylor, MdSA.

Stoney. Anne Arundel County, Md., 1773, 1 acre, Joseph Jacobs, MdSA.

Stoney. Cecil County, Md., 1903, 1 acre, 1 perch, Frederick W. Irwin, MdSA.

Stratton. Between Long Creek and Battsses Bay in Lynhaven, Va., 1711, 238 acres, Edward Attwood, Nugent, vol. 3.

St. Thomas. Somerset County, Md., 1821, 152.75 acres, Thomas Jones, MdSA.

Sudlers. Queen Anne's County, Md., 1743, 64 acres, Joseph Sudler, MdSA.

Sugar. Harford County, Md., 1804, 82 acres, Joseph Scarborough et al., MdSA.

Sunken. Dorchester County, Md., 1759, 10 acres, John Staples, MdSA.

Sunken. Dorchester County, Md., 1902, 117 acres, 19 perches, James D. and Lawrence Slacum, MdSA.

Swamp. St. Mary's County, Md., 1795, MdSA.

Swan. Anne Arundel County, Md., 1810, 2.375 acres, Wm., John, and Elizabeth Gwynn, MdSA.

Swan. Kent County, Md., 1687, 16 acres, William Frisby, MdSA.

Swan. Somerset County, Md., 1958, 28.77 acres, H. Harvey Bradshaw, MdSA.

Swan. Somerset County, Md., 6.4 acres, part of Smith Island area, 1 mi. NW Ewell, Gelenter.

Terrapin. Anne Arundel County, Md., 1774, 5 acres, Samuel Lane, MdSA.

Terrapin. Anne Arundel County, Md., 1789, 20.25 acres, Samuel Lane, MdSA.

Terrapin Sand. Somerset County, Md., 1773, 140 acres, Benjamin and Robert F. Leach, MdSA.

Thorny Point. Prince Georges County, Va., 1729, 324 acres, part of islands near Tangier Island, John Gillum, Nugent, vol. 3.

Three Partners. Cecil County, Md., 1818, 77 perches, Isaac Hawkins, MdSA.

Trail Fair. Harford County, Md., 1834, 2 acres, John Kirk, MdSA.

Triangle. Somerset County, Md., 1943, 37 rods, M. Morris Whitehurst, MdSA.

Troy. Somerset County, Md., 1959, 4.59 acres, H. Harvey Bradshaw, MdSA.

Turkey. Anne Arundel County, Md., 1696, 333 acres, Neale Clarke, MdSA.

Turkey. S. side of Meherrin River, Brunswick County, Va., 1726, 100 acres, Ralph Jackson, Nugent, vol. 3.

Turkey. Harford County, Md., 1819, 7 acres, John Kirk, MdSA.

Turkey Buzzard. Calvert County, Md., HCaC.

Turkie. Henrico County, Va., 1639, 300 acres, Richard Cocke, Nugent, vol. 1.

Turtle Egg. In Holland Straits, 1 mi. NE of Deep Banks Island, Dorchester County, Md., 15.9 acres, Gelenter.

Vicaris. In Rappahannock River, Essex County, Va., 1720, Francis Thornton, Nugent, vol. 3.

Vinsons. Dorchester County, Md., 1734, 10 acres, James Vinson, MdSA.

Wards. Somerset County, Md., 1923, 33.25 acres, Eugene H. Johnson, MdSA.

Watts. S. of Watts Great Island and Gabriells Island, Accomac County, Va., 1675, 34 acres, John Taylor, Nugent, vol. 2.

Watt Taylor's. Charles City County, Va., 1670, 400 acres, Ambrose White, Nugent, vol. 2.

Weed. Harford County, Md., 1836, 3.25 acres, Henry Wesley, MdSA.

West Troy. Somerset County, Md., 1959, .41 acre, H. Harvey Bradshaw, MdSA.

Whimwham. In N. Poquoson River, Princess Anne County, Va., 54 acres, William Wood, Nugent, vol. 3.

Whitehurst. Somerset County, Md., 1939, 12.8 rods, M. Morris Whitehurst, MdSA.

Whitehursts. Kent County, Md., 1922, .145 acre, Ruth B. Whitehurst, MdSA.

White Rock. Cecil County, Md., 1818, 1 rod, 15 perches, Joseph Miles et al., MdSA.

Whites. Lower Norfolk County, Va., 2,600 acres, Col. Lemuel Mason, Thomas Jarvis, and Thomas Willoughby, Nugent, vol. 2.

Wild Cat. Harford County, Md., 1818, 27 perches, John Kirk, MdSA.

Wild Grove. Cecil County, Md., 1817, 3 acres, 1 rod, 33 perches, Reuben Reynolds, MdSA.

Willeys (formerly Hambletons). Two mi. S of St. Michaels, Broad Creek, Md., MdSA.

Wills. Upper side of Flat Creek, Prince Georges County, Va., 400 acres, John Burton, Nugent, vol. 3.

Winters. Harford County, Md., 1839, 10 acres, John Donovan et al., MdSA.

Woody. Between Pungoteague and Anoncock Creeks, Va., 1698, 9.5 acres, Thomas Taylor, Nugent, vol. 3.

Woodyard. Charles County, Md., 1740, 40 acres, Thomas Posey, MdSA.

Yieldhalls. Anne Arundel County, Md., 1679, 100 acres, Wm. Yieldhall, MdSA.

Yorkshire. Somerset County, Md., 1664, 100 acres, William Wilkinson, MdSA.

## UNNAMED ISLANDS

1624, James City County, Va., 12 acres, Wm. Spencer, Nugent, vol. 1.

1627, James City County, Va., 200 acres, Adam Dixon, Nugent, vol. 1.

1628, James City County, Va., 200 acres, Isabella Perry, Nugent, vol. 1.

1628, James City County, Va., 10 acres, Robert Marshall, Nugent, vol. 1.

1637, James City County, Va., 600 acres, Benjamin Harrison, Nugent, vol. 1.

1638, Ely County, Va., 24 poles sq., Wm. Parry, Nugent, vol. 1.

1638, James County, Va., 10 poles by 8 poles, Derrick and Arent Cortsen Stam, Nugent, vol. 1.

1638, James County, Va., 6 poles by 4 poles, Wm. Parry, Nugent, vol. 1.

1641, Charles City County, Va., 250 acres, on land of Capt. Epps, Walter Astoy, Nugent, vol. 1.

1643, James City County, Va., 1 acre, John Watson, Nugent, vol. 1.

1644, James County, Va., 8 acres, Thomas Hampton, Nugent, vol. 1.

1649, York County, Va., 1,000 acres joining Warranuncock Island, N. side of York River, Nicholas Jernew, Nugent, vol. 1.

1652, James City County, Va., 196 acres, Edward Travis, Nugent, vol. 1.

1652, on NW branch of Nansemond River, Nansemond County, Va., 100 acres, Henry and Elizabeth Lawrence, Nugent, vol. 2.

1656, James County, Va., 15 acres, Jno. Baldwin, Nugent, vol. 1.

1657, James City County, Va., 100 acres, Thomas Woodhouse and Wm. Hooker, Nugent, vol. 1.

1657, James City County, Va., 150 acres, Richard James, Nugent, vol. 1.

1658, James City County, Va., .5 acre, Wm. Harris, Nugent, vol. 1.

1664, Lower Norfolk County, Va., 100 acres, Wm. Broch, Nugent, vol. 1.

1668, by Filgates Island, near James River, Isle of Wight County, Va., 750 acres, Thomas and Mary Bland, Nugent, vol. 2.

1672, against Dumplins Island on Troublesome Point, on Nansemond River, Nansemond County, Va., 350 acres, Thomas Milner, Nugent, vol. 2.

1678, James County, Va., 6 poles by 4 poles, Wm. Paray, Nugent, vol. 1.

1679, on W. side of Highs Island, Isle of Wight County, Va., 200 acres, Robert King, Nugent, vol. 2.

1682, near Capt. Underhills Island, S. side of river upon Chickahominy main, New Kent County, Va., 140 acres, Nugent, vol. 2.

1683, Nansemond County, Va., 150 acres including Rackoone Island, upper part of Nansemond River, Thomas Parker, Nugent, vol. 2.

1687, Gloucester County, Va., 25 acres, an island adjacent to Oak Island and with other small islands, Edward Mumford, Nugent, vol. 2.

1687, in Reedy and Powhite Creeks, Henrico County, Va., 5,075 acres, includes Harwood, My Lords, Prices Folly, and Willows Islands, William Byrd, Nugent, vol. 2.

1713, Northwest River, Va., 3 islands, 17 acres, Henry Horstead, Nugent, vol. 3.

1714, a little above the falls of the Pamunkey River, King William County, Va., part of 400 acres, James Terry, Nugent, vol. 3.

1714, Falls of Pamunkey River, Va., 400 acres including island, William Fleming, Nugent, vol. 3.

1714, part of 245 acres, S. side of W. branch of Nansemond River, John Meredith, Nugent, vol. 3.

1717, Prince Georges County, Va., 431 acres including an island, beginning N. side of Nottoway River, John Freeman, Nugent, vol. 3.

1720, Essex County, Va., 41 acres including an island, Rowland Thornton, Nugent, vol. 3.

1722, Surrey County, Va., part of 490 acres S. side of Meherrin River, Samuel Sanford and Daniel Crawley, Nugent, vol. 3.

1772, island in Chickahominy Swamp, 145 poles below Timber Bridge, Charles City County, Va., Thomas Flowers Nugent, vol. 3.

# INDEX

Harrison, John B., 73; boat builder, 72
Hart, Joseph, 30
Hart and Miller Islands: dredge spoil containment area, 30; North Point State Park, 31
Hawkins, Thomas, 57
Helmsley, Alexander, 70
Henderson, Gilbert, 103
Henry, Clement, and Sitka deer, 76
herons, 6, 40; on Barren Island, 93; on Bloodsworth Island, 95; on Bodkin Island, 51; on Holland Island, 100; on Poplar Island, 60; on Smith Island, 132; on Watts Island, 103
*Hester Ann* (sloop), lost in storm, 58
Hickory Cove, steamboat wharf, 87
Hole in the Wall, at mouth of Milford Haven, 151
Holland, Daniel, 95
Holland Island: environmental refugees from, 4; housing on, 94; hunting lodge on, 100; land loss, 6, *96;* life on, 6, 97; Methodist Episcopal Church, 96; storm of 1918 and, 6; vegetation on, 100
Holland Island Bar Lighthouse: on Kedges Strait, 96; navy bomber attack on, 96
Holland Straits, 6, 95; cemetery, 6
Honga, Hoopers Island, 83
Hooper, Henry, 84
Hooper, Henry, Jr., 84
Hooper, William, 84
Hoopers Island: bridge, 87; British raids on, 85; economy, 88, 95; land loss on, 88; lighthouse, 87; Memorial Church, 87: windmill, 82
Hoopersville, 87
Hooten, Thomas, indentured servant on Hoopers Island, 84
Hopkins, George F., indentured servant on Holland Island, 96

Hopkins, William, 93
Horseshoe Point, Three Sisters Islands, 160
Horton, Tom, 130
House Islet, and Great Fox Island group, 101
Howgate, H. W., 143
Hughes, Samuel, 23
Hunter, J. F., erosion study, 7
hunting: on Barren Island, 89, 90; on Bloodsworth Island, 94; on Deal Island, 111; on Eastern Neck Island, 43; on Gibson Island, 35; on Holland Island, 97; on James Island, 76; methods, 12; pheasant, 55; on Pooles Island, 28; on Poplar Island, 59; on Spesutie Island, 24; on Susquehanna Flats, 12; on Taylors Island, 82; on Tilghman Island, 73; on Wye Island, 63
hurricane damage: to Cobb Island, 150; to Deal Island, 112; to Janes Island, 125; to St. George's Island, 146; to Tangier Island, 137, 141; to Watts Island, 106
Hurricane Hazel effect: on Cobb lsland, 150; on St. George's Island, 146
Hurricane Isabel effect: on Broomes Island, 9; on Cobb Island, 9; on Fox Island, 8; on Havre de Grace, 9; on Hoopers Island, 8; on Kent Island, 9; on St. Michaels, 8; on Tangier Island, 8; on Watts Island, 8, 106
Hyer, Tom, boxing match on Pooles Island, 28
Hynson, Thomas, 43

Ice Age, Chesapeake Bay and, 6
*Ida May* (skipjack), 111
indentured servants, 56–57, 84; Africans as, 156–57
Indians. *See individual tribes*
Ingleside, Eastern Neck Island, 43
Iroquois, 16

*Island Belle I,* and Smith Island, 134
Island Point, Janes Island, 124

J. C. Lore & Sons, on Solomons Island, *116, 117*
*J. T. Leonard* (sloop), 73, *75,* 76, 82
Jacobus Creek, Cacaway Island, 44
Jacques, Willard, 55
James, Henry, 18, 19
James Fort, Jamestown Island, 158
Janes Island: land loss on, 125; lighthouse, 124; light ship, 124; plank road on, *75;* state park, 123, 125
Jefferson Island, *58, 59;* former Cobbler's Neck, 59
Johnson, C. Lowndes, of Gibson Island Yacht Squadron, 36
Johnston, J. Edward, 42
Jones, David, 43

Kedges/Cager's Strait, 96, 101
Keene, Benjamin, 80
Keene, Harry, 80
Kent Fort, Kent Island, 45
Kent Island: development on, 49; fur trade, 13, 16; historical society, 45, 48; Kent Narrows, 49, *50;* land loss on, 49; William Claiborne and, 45, 46
Kent Point, Kent Island, 45
Kidd, John, and Battle of Kedges Strait, 101
King, Robert, 103
Kleinfelter, G. Randolph, 103, 106, 142
Knapp, Robert, 69
Knapps Narrows, 69, 70, 71

land loss. *See individual islands*
Lane Methodist Episcopal Church, Taylors Island, 82
Langford, John, 44
Laws, John, 107
least terns: on Assateague Island, 92; on Barren Island, 91, 92

Lee, Robert E., and Fort Carroll, 32
Leonard, James T., 73
Lewis, Margaret, 144
Little Island, and Dobbins Island, 38
Little Watts, 104, 106
Lloyd, Edward, 65
Lloyd, John Howard, 65
Lloyd, Margaret, wife of Matthew Tilghman Ward, 70
Lloyd, Philemon, 61
Lloyd family, 61
log canoe/dugout canoe, 68, 72
Long Island, and Holland Island, 95
Long Island (New York), Battle of, 80
*Long Tayle,* and William Claiborne, 46
Lord Baltimore, Cecil Calvert, 13, 16, 23, 46, 153
Love Point, Kent Island, 48, 49
Lowery Seafood Company, Broomes Island, 120
loyalists (Tories), and battle of the barges, 136; on Broomes Island, 119; on Deal Island, 107; on Devil's Island, 108; on Eastern Shore, 14; on Great Fox Island group, 101; on Gwynn's Island, 151; on Hog Island, 103; and Holland Island, 95; on Hoopers Island, 85; and Lord Dunmore, 128; and Lower Bay islands, 128, 129; and Middle Bay islands, 68; on Spring Island, 95: on St. George's Island, 129; and Tangier Island, 103, 136; and Tilghman Island, 70; on Watts Island, 103
Luckenbach, Al, Anne Arundel County archaeologist, 38
lumber industry: and Maids Island, 19; and Taylors Island, 82
Lusby, Clifford, oyster tonger, *115*
*Lydia Louise I, II* (research vessels), vi
Lynch, Stephen, 143

Quomacac, 118

radio-telephone, on Cobb
   Island, 129, 150
Raney, Eugene, 40
Ray, Thomas, 65
*Reprisal*, Lambert Wicks
   of, 42
Revolutionary War. *See under
   individual islands*
Rickets, Samuel, 160
Ridgeway, R. O., 58
Ringgold, James, 53
Ringgold, Mary, 53
Robrect, John, 145
rock fish, 20, 21
Rogers, Mary, 97
Rolfe, John, and tobacco
   cultivation, 156
Romancoke, Kent Island, *48*
Roosevelt, Franklin Delano,
   and Jefferson Island Club,
   *58, 59*
Rozelle, Ennis, 145
Rubentown, Tangier Island, 137
Russel, Douglas, Oyster Navy
   captain, 148
Russell, Walter, 130
Russels Isles: John Smith and,
   130, 136; and Smith and
   Tangier Islands, 130, 136
Ryan, Thomas, 53

sailmaking, 111
salt marsh, 6, 7, 92, 94, 95,
   100, 130, 135, 142
Sappington, Benjamin, 43
Sappington, Frances Wickes,
   43
sawmill(s), 19, 70, 82, 145
Schock, Mildred C., 45
schooners, 72, 82, 115
Schubel, Jerry, vi, vii
Schuyler, Sidney, and Bruff's
   Island causeway, 65
seafood industry; on Broomes
   Island, 120; on Cobb Island,
   150; on Deal Island, 107; on
   Hoopers Island, 88; on
   Kent Island, 48; on Smith
   Island, 132; on Solomons Island, 115; on St. George's Island, 146; on Tangier Island,
   137; on Taylors Island, 82

sea level rise: on Bloodsworth
   Island, 95; effect on islands,
   4–8; on James Island, 76;
   94; 95, 101, on Smith Is-
   land, 135; on Tangier Island,
   140; on Watts Island, 103
Sears, William, 59
seine fishing, 17, 20, 51, 120
Seneca Indians, 16
Seven Island Duck Club, on
   Bodkin Island, 51
Severn, Battle of the, 113
Sewell, C. R., and seine
   fishing, 120
shad fishery, 13, *17*, 20, 21, 97
Shanks Hammock, Black-
   beard and, 141
Shapeley, Philip, 84
Sharp, Peter,161
Sharps Island: erosion rate, 7;
   farms, *162;* hotel, 162; light-
   houses, 162–63; and War of
   1812, 161
Shaw Bay, 64
Sherwood, John, and Gibson
   Island Yacht Squadron, 37
Sherwood, William, 65
ship wreck, off Tangier
   Island, 141
Shomette, Donald G., 77, 103
silk worms, on Taylors
   Island, 78
Silver, Charles, 17
Simmons, Raymond, Jr., 92
sink box gunners, 12, 13
Siskin, William, 90
Sitka deer: on Assateague Is-
   land, 69; on James Island,
   *75, 76*
skipjack(s), *132;* Deal Island,
   *111;* and oyster fleet, 72, 111;
   races, 111; sloop-rigged, *75*
Slaughter Creek, Taylors Is-
   land: bridges, 82; broads, 77
Slaughter Narrows, and Taylors Island, 77
slaves/slavery: Bordley and,
   62; British prisoner, 81;
   Methodists and, 82; on
   Sharps Island, 161
sloops, built: on Taylors
   Island, 82; on Tilghman
   Island, 72
smallpox, 129; on Gwynn's

Island, 151; on Tangier
   Island, 137
Smith, Henry, 130
Smith, John, 4; exploration,
   12; and Garrett Island, 15;
   and Indian villages, 118; and
   Jamestown, 155–56; Ozinies,
   41; and Pooles Island, 27;
   and Russels Isles, 139;
   Tilghman Island, 69
Smith, John D., 24
Smith, Robert, 24
Smith, Ruth Mackall, oyster
   shucker, 120
Smith, William, 24
Smith Island, 8; growth on,
   131; land loss on, 130;
   population decline, 134;
   museum, 134
Smithsonian Institution, and
   Poplar Island, 59
*Smokey Joe,* and steamer
   *Philadelphia,* 49
Snyder, Charles, 53
soft-shell crab industry, Deal
   Island, 112
Sollers Point Flats, Fort
   Carroll, 32
Solomon, Isaac, fishing and
   canning operations, 115
Solomons Island: and Molly's
   Leg Island, 116; and World
   War II, 117
Solomons Marine Hospital,
   cemetery on Molly's Leg
   Island, 116
*Somerset* (skipjack), 111
Somervell, Alexander, 113
Somervell, James, 113
South Little Fox Islet, part of
   Great Fox Island group, 101
Spartina grass, 7
Spencer, Richard, II, 43
Spencer Hall, Eastern Neck
   Island, 43
Spesutie Island: Aberdeen
   Proving Ground, 25; and
   British raiders, 14; erosion
   rate, 25; manor ghost, 24;
   hunting lodge, 24; steam-
   ship coaling station, 13; and
   Susquehannocks, 22; and
   World War I, 24
Spesutie Narrows: causeway,

25; ferry, 24; winter crossing, 24
Spicer, Jeremiah, 81
spoil island, Fishing Battery
   Island, 21
Spring Island, British on,
   68, 95
Spry, Oliver, 160
Stansby, Mary, 160
Staplefort, Raymond, 78
Star boats, of Gibson Island
   Yacht Squadron, 37
St. Clement's Island: devel-
   opment on, 154; fort, 8;
   land loss, 8; lighthouse,
   153–54; Potomac Museum,
   154; resort, 154; state
   park, 154
steamboat(s): to Deal Island,
   112; to Gwynn's Island, 152;
   to Hoopers Island, 87; to
   Kent Island, 48; to St.
   Clement's Island, 154; to
   Sharps Island, 162; travel,
   13, 71–72, 87, 152
Steelpone Creek, Cacaway
   Island, 44
Stevens, William, 61
Stevensville, Kent Island, 48
Stewart, Anthony, and *Peggy
   Stewart,* 161
Stewart, Glenn, 63
Stewart, James, and British, 81
St. George's Island: bridges,
   146; development, 146; and
   Lord Dunmore, 129; re-
   sort, 146; and Revolution-
   ary War, 145; shipyards, 145;
   and War of 1812, 145
St. Helena Island: gambling
   casino, 40; handicapped
   children's camp, 39; Indians
   on, 40; mansion, 39
St. Inigoes Manor, Jesuits, 145
St. John's Creek, windmill
   on, 82
St. John's United Methodist
   Church, Deal Island, 111
St. Mary's City, William Clai-
   borne and, 13
St. Paul's Catholic Church,
   Cobb Island, 150
Stubbs, John, 111
Stump, John W., 18